Rethinking Energy: Ripple Tension, Not Substance

Sandeep Chavan

Published by SANDEEP CHAVAN, 2025.

RETHINKING ENERGY: RIPPLE TENSION, NOT SUBSTANCE

First edition. August 16, 2025.

ISBN: 979-8231100590

Written by Sandeep Chavan.

Table of Contents

Chapter 1: The Birth of a Concept 1
Chapter 2: Einstein's Equation – Mass Hysteria 14
Chapter 3: Quantum Confusion 31
Interlude – Leaving the Illusion Behind 56
Part II: UED – Energy Redefined 58
Chapter 4: The Real Nature of Energy 59
Chapter 5: Dissolution = Release 86
Chapter 6: Electricity – The Ripple Tension Game 112
Interlude – From Belief to Behavior 138
Part III: Technological Rethink 141
Chapter 7: Machines That Mediate 142
Chapter 8: Batteries and Storage 164
Chapter 9: Power Grids and the Illusion of Flow 184
Interlude: Part III – Technological Rethink 209
Part IV: New Frontiers 211
Chapter 10: Fusion, Fission, and Fallout 212
Chapter 11: Renewables, But Real 230
Chapter 12: Quantum Devices and Ripple Precision 248
Interlude – We Never Controlled Energy 269
Part V: Implications and Applications 271
Chapter 13: Educational Reset 272
Chapter 14: Policy, Climate, and the Myth of Energy Crisis. 277
Chapter 15: Toward a Ripple-Based Civilization 282
Epilogue – The Field Remembers 288
Appendices 290
Appendix A – UED Equations 291
Appendix B – Glossary of Core UED Terms 296
Appendix C – UED vs Classical Energy Models 300
Appendix D – Suggested Classroom Experiments 305
Acknowledgments 310
About the Author 312

Bold Declarations of Universal Energy Dynamics (UED) 314

To the pioneers of the energy phenomenon —from Faraday to Feynman, Tesla to Einstein, Planck to Bohr —who lit the first spark in our understanding of power, force, and motion.

This book stands not in defiance,but in deep gratitude.

For without your questions,we would have never dared ask a different one.

May this ripple carry your legacy forward — not as correction, but as continuation.

"The world was never powered.It was permitted.In stillness, it strained.In tension, it waited.And when alignment came,it moved —not as force,but as grace."

— *Chavan*

Disclaimer

This book is a work of independent scientific philosophy and reinterpretation. It challenges conventional physics, engineering, and educational models using the lens of Universal Energy Dynamics (UED) — a framework developed outside traditional academic institutions.

While the ideas presented here are grounded in logical reasoning and experiential consistency, they do not conform to currently accepted mainstream theories or textbook definitions.

The content is intended for critical thinkers, educators, researchers, and students who are open to exploring alternative perspectives on energy, motion, and causality. It should not be mistaken for official scientific consensus or used in place of certified academic curricula unless supported by appropriate discussion and guidance.

The author is not affiliated with any institution or governing scientific body, and the views expressed are solely his own. Readers are encouraged to examine these ideas thoughtfully, skeptically, and independently — and to validate insights through direct observation, dialogue, and lived experimentation.

This book is not here to persuade. It is here to provoke alignment — if and only when your field is ready.

Opening Note

By Sandeep J. Chavan

For centuries, we believed we were harnessing energy—storing it, converting it, even weaponizing it. We built our civilization around this belief. But what if I told you that energy was never what we thought it was? That it doesn't exist as a substance, but as a symptom—a ripple tension arising from the universe's silent struggle to return to balance?

This book is not just a critique of modern science. It is a complete deconstruction of one of its most sacred assumptions. Through the lens of Universal Energy Dynamics (UED), we will see that nothing flows—not electricity, not light, not even time. What we observe as "energy transfer" is merely the universe resolving its discomfort.

Every machine, every explosion, every beam of light is not an act of energy in motion, but ripple misalignment seeking relief. From the battery in your remote to the turbines spinning in a nuclear plant—they are all mediators, not creators.

This isn't just about physics. It's about honesty. It's about having the courage to challenge a model that works *enough* to be useful, but not enough to be true. If you've ever felt there's something off with the way we define power, force, motion, and matter—you were right to question.

This book is for scientists, engineers, philosophers, educators, and curious minds who are done with patchwork explanations. It is for those who are no longer satisfied with watching the world perform but want to understand the stage on which everything occurs.

Let us rethink energy—not as a unit, but as a consequence. Not as a source, but as a symptom. Not as a possession, but as a temporary imbalance.

The revolution begins at the very wire you thought carried power. It doesn't.

Preface

"This is not a book about energy. It's a book about what we misunderstood while chasing it."

We were taught that energy flows, that it powers the world, that it can be stored, transferred, consumed.

We were handed equations, diagrams, and metaphors — from bouncing balls to burning fuels, from electrons in wires to photons in space. And we believed them. Because they worked. Or at least, they worked well enough to light our homes, run our machines, and launch our rockets.

But somewhere between what we measured and what we truly understood, something was lost. A silence. A stillness. A deeper causality behind all that seemed to move.

This book is the result of years of listening to that silence — as an engineer, a teacher, a physicist, and most importantly, an observer.

It challenges the most fundamental beliefs we hold about energy — not with rebellion, but with clarity. Not with complexity, but with coherence.

It doesn't aim to rewrite textbooks. It aims to reveal what the textbooks never told us: That energy is not a thing. It is a resolution. That nothing flows. Everything aligns — or waits to.

Through this work, I invite you to see the universe as a ripple field — where tension, not mass, drives events. Where light, heat, electricity, and gravity are not mysterious forces, but patterns of permission. Where intelligence isn't coded into machines, but emerges wherever ripple alignment becomes self-aware.

This is not an academic theory. It is a lens. A lens you can carry into your classroom, your lab, your city, your mind — and never see energy the same way again.

If it resonates with you, it will not be because you agree with everything here. It will be because you felt something shift. A ripple aligned. A misalignment exposed. A question no longer afraid to be asked.

That's all this book hopes to do — *not give answers,* but restore the permission to wonder, and perhaps... to rethink everything we thought we knew about energy.

— Sandeep J. Chavan

Reading Guide

"Don't rush to understand. First, learn to see."

Who is this book for?

This book is written for:

- Students who feel something is missing in what they're taught about energy.
- Teachers looking to spark new conversations in the classroom.
- Scientists and engineers willing to question deeply-held assumptions.
- Philosophers and thinkers exploring the border between mind, matter, and motion.
- Everyday observers who have quietly sensed: *"It doesn't really work the way they say it does."*

No prior expertise in physics is required — only curiosity, humility, and a willingness to unlearn.

How should you read this book?

This is not a textbook. It doesn't build linearly like a syllabus — it spirals, like a ripple.

Each part peels away layers of illusion. You may revisit earlier chapters with new eyes after finishing later ones. That's expected.

Here's one suggested flow:

Part	Focus
Part I: Energy as Belief – The Historical Illusion	Deconstructing what "energy" is not. Unpacking myths.
Part II: UED – Energy Redefined	Introducing ripple dynamics, ATT, and the logic of alignment.
Part III: Technological Rethink	Reframing devices, grids, and observed phenomena.
Part IV: New Frontiers	Applying ripple theory to nuclear, renewable, and quantum systems.
Part V: Implications and Applications	Education, policy, civilization — the deeper ripple.
Appendices	Practical tools, glossary, classroom experiments, and declarations.

Tips for Getting the Most Out of It

- Pause often. This isn't fast reading — it's field reading. Let ideas settle.
- Journal ripple moments. If something "clicks" or disturbs you, note it. That's realignment happening.
- Discuss openly. Many of these ideas are meant to be challenged — use them to spark dialogue in your circles.
- Don't force agreement. Let ideas pass through you. The field will decide what resolves.

For Teachers and Facilitators

You'll find suggested experiments, discussion prompts, and glossary entries in the appendices. Encourage students not to memorize—but to observe, sense, and question alignment and misalignment in the real world.

Every classroom can be a ripple lab.

What Comes After This Book?

This book is the entry point. The permission slip.

If you feel the ripple, you're already inside the field of UED.

More explorations await — into consciousness, the human system, intelligence, death, time, and beyond — all through the ripple lens.

But start here. Start still.

"Before we moved the world, we should have asked — did the world allow it?"

Prologue: The Lie We Lived

"There was no flowing force. Only unresolved tension."

We lived a lie. Not because we wanted to. Not because we were ignorant. But because it *worked*. The concept of energy—so elegant, so useful—fit the times, the tools, and the minds that forged it.

It became our god in equations, our ghost in machines. We worshipped it. Measured it. Chased it. Traded it. But never truly understood it.

The Cult of Energy

From steam engines to smartphones, we declared that *energy makes everything go.* It became the invisible fuel behind all change, all power, all action. Energy was no longer a phenomenon—it was a belief system. A cult wrapped in the lab coat of science.

Children are taught the "forms of energy" as if listing pantheons of gods: kinetic, potential, thermal, electrical, chemical, nuclear. We assign units to it—joules, kilowatt-hours—and pretend we've captured its essence. But what if these forms are just shadows, flickers of ripple tension—not entities in themselves?

The lie wasn't just in what we believed. It was in how deeply we embedded it into our reality, never once questioning its foundation.

Why $E = mc^2$ Misled Us

Einstein gave us a symbol so powerful it became scripture: $E = mc^2$.

But symbols are only as good as their assumptions. This equation—revolutionary and beautiful—was never the truth, only a *symmetry between observations* under extreme conditions. It seduced generations into believing that mass contains energy, and that energy is the ultimate essence of the universe.

But under UED, we now see: Mass is ripple configuration. Energy is ripple tension. Neither is an object. Both are outcomes. Mass doesn't "convert" into energy. It resolves into a more relaxed state of the structured vacuum field. What we call "energy release" is not liberation—it is realignment.

From Force to Fiction: How Energy Became the Currency of Science

Science once sought *causes*. But the moment energy became central, it began chasing *effects*. We stopped asking *why* and started measuring *how much*.

Force, power, work, even heat—all became transactions of energy. The universe was rebranded as a marketplace of flows and exchanges, ignoring that beneath all of it, there was no substance moving. No carrier. No flow.

Only ripple tension.

The real currency of reality isn't energy—it is imbalance resolving. But we mistook every flicker of restoration as a movement of "stuff." And in doing so, we built our entire physics on a practical illusion.

It's time to correct the course. It's time to dissolve the cult and restore science's original pursuit: truth over convenience. We begin by rethinking the very word we've all grown to trust—energy.

Part I: Energy as Belief – The Historical Illusion

Energy was never a "thing." It was a mathematical placeholder—a tool created to close equations, not to describe a physical reality.

Chapter 1: The Birth of a Concept

"It was never discovered. It was invented. And then we forgot we invented it."

Energy—the cornerstone of modern physics, the currency of machines, the god of equations. We speak of it as if it flows through wires, bursts from stars, hides inside atoms, and powers our thoughts. But step back, and ask: what is it really?

Not one scientist has ever seen energy. We have only seen change. Movement. Heat. Rotation. Explosion. And to make sense of it all—to measure, to predict, to control—we invented a word. A number. A bookkeeping tool.

Energy.

Born in the coal furnaces of the Industrial Age. Raised by steam engines. Adopted by thermodynamics. Worshipped by physicists.

But here's the problem: somewhere along the way, we mistook the map for the territory. The abstraction became the explanation. And now, two centuries later, our entire understanding of the universe rests on a concept that was never physically real.

This is the story of that concept. How it was born. Why it misled us. And how, under UED, we can finally set the record straight.

Let us begin not with the universe—but with the steam engine.

1.1 The Industrial Spark – Steam, Heat, and Work

The story of energy does not begin with the stars or atoms—it begins with steam.

The 18th and 19th centuries marked a turning point in human history. Coal was being pulled from the earth by the ton, factories were rising across Europe, and machines—those clanking, hissing marvels—were suddenly doing the work of men and horses. This was the dawn of the Industrial Revolution, and with it came a practical, urgent question:

How much work can a machine do?

It was not a philosophical inquiry. It was a problem of profit and performance. Entrepreneurs and engineers needed to know whether a new engine was worth the coal it burned. A new language had to be invented—not to explain nature, but to serve industry.

Enter James Watt, whose refinements to the steam engine revolutionized mechanical power. But Watt was not just a tinkerer. He also introduced the concept of "horsepower"—a metric for comparing engine output to the labor of a horse. This was the birth of mechanical quantification.

Then came Sadi Carnot, a French physicist who asked a deeper question: *Why are some engines more efficient than others?* In trying to answer it, Carnot laid the foundations of thermodynamics, even though the term "energy" was still not clearly defined. His key insight: heat seemed to flow from hot to cold and could somehow be converted into mechanical motion—but never entirely. Some of it was always "lost."

And so arose a puzzle: If machines convert heat into motion, then what exactly is being transferred? What is being lost or preserved? What is the common element behind heat, motion, lifting weights, turning gears?

To make sense of this, physicists introduced the concept of "work"—defined as *force times displacement*. Push an object with a certain force over a certain distance, and you have "done work." But still, this was just a mathematical relation, a description of movement—not a substance.

Eventually, the term energy entered the scene—first in mechanical contexts, then thermodynamic, and finally generalized across physics. It was never seen, only inferred. But it made the numbers balance.

It told you how much coal you needed. It told you how much heat was lost. It helped design better machines.

But make no mistake—energy was not discovered in nature. It was invented in the factory.

It served the engineer long before it seduced the scientist. It became an abstraction for measuring and optimizing machine performance, not explaining reality.

Key Insight: Energy was not born to describe the universe—it was born to predict how well a steam engine would run.

And yet, from that humble, practical origin, a new belief was born: that everything runs on energy.

But before we reach that dogma, we must understand how this abstraction took hold in science.

1.2 Energy as a Bookkeeper – The Accountant's Tool

By the mid-19th century, science was undergoing a transformation—not driven by wonder, but by utility. The age of speculation was giving way to the age of bookkeeping, where every input and output in a system had to be accounted for—not in gold or goods, but in something far more abstract: energy.

At the heart of this shift was Rudolf Clausius, a physicist and mathematician who played a central role in formalizing thermodynamics. Clausius didn't think of energy as a mysterious cosmic fluid or a divine spark. He thought of it like an accountant thinks of money: something to be conserved, transferred, debited, or credited, depending on the process.

What earlier thinkers had called "caloric"—a supposed heat substance—was now obsolete. Clausius reframed it: heat is not a substance but a mode of energy transfer. This led to the formulation of the First Law of Thermodynamics, which stated:

Energy cannot be created or destroyed. It can only change form.

But what were these forms?

Suddenly, the scientific vocabulary expanded:

- **Kinetic energy**: the energy of motion.
- **Potential energy**: the energy of position or configuration.
- **Internal energy**: the sum of all molecular movements and interactions inside a system.

None of these were observable in any direct way. You couldn't touch potential energy. You couldn't see internal energy. But the math worked. When you assumed these quantities existed, equations balanced. Predictions came true. Machines performed as expected.

And so, energy became the placeholder that made the numbers make sense. It didn't need to be *real*—just reliable. It didn't need to be *understood*—just useful.

It was like a currency in a complex economy. You didn't need to see the gold in the vault; you just needed the books to balance. The conservation of energy became the Golden Ledger of physics, the universal law that nothing is gained or lost—only moved, converted, or dissipated.

Whether a falling rock, a spinning wheel, or a burning log—energy was there. Not as an object, but as a value assigned to each action, event, or transformation.

This mathematical abstraction, though invisible, became indispensable. It powered predictions. It improved engines. It justified funding. And over time, we forgot what it really was: A bookkeeper's invention, not a discovery of nature.

We didn't care. The equations worked. The machines ran. That was enough.

Core Argument: Energy was never a "thing." It was a mathematical placeholder—a tool created to close equations, not to describe a physical reality.

In our eagerness to engineer the world, we had accepted an illusion because it was efficient.

But illusions—especially successful ones—have a way of becoming dogma.

1.3 The Seduction of Success – When Approximation Becomes Belief

What began as a mathematical trick soon became the bedrock of belief. The equations were clean, the results matched experiments, and the industrial world was booming with innovation. In such a climate, nobody paused to ask whether the abstraction they were using—energy—was actually real.

The success of the energy framework was so overwhelming that it produced a kind of philosophical laziness. The original intent had been clear: energy was a computational construct, a conservation tool, a placeholder to track transformations.

But the more it worked, the less we questioned it.

Instead of saying: "Energy helps us balance the math," we began to say: "Energy makes things happen."

A subtle yet dangerous linguistic shift had occurred. Equations became explanations. Approximation became ontology—the study of what really exists.

We stopped asking:

- *What is energy, really?*
- *Where does it reside?*
- *Can we isolate it, see it, touch it, measure it directly?*

Instead, we invented phrases that quietly solidified the illusion:

- "Energy is stored in a battery."
- "Energy is released during a chemical reaction."
- "Energy flows through a circuit."

But these are metaphors—not measurements.

You cannot hold "stored energy" in your hand. You cannot bottle up a "flow" of energy like you can water. You cannot extract energy as a substance from a machine.

Still, the language sank into textbooks, engineering manuals, and popular science. By the early 20th century, the transformation was complete: Energy was no longer a tool—it was the actor, the entity, the substance behind everything.

Engineers treated it as fuel. Textbooks taught it as truth. Culture absorbed it as fact. And the illusion grew stronger with each generation that used it without questioning it. What worked became what was real. Success had seduced us into belief.

Ripple View Insert — UED Perspective

From the lens of Universal Energy Dynamics (UED), all these so-called "forms of energy" are not fundamental entities, but merely ripple tensions resolving across different scales and alignments.

- **Kinetic energy**? Ripple alignment manifesting in

directional coherence.

- **Potential energy**? Ripple tension accumulated in an off-equilibrium configuration.
- **Thermal energy**? Random micro-ripples seeking localized balance.
- **Internal energy**? The sum of distributed ripple stress in a system.

In truth, there is no "thing" called energy moving or being stored. There is only Structured Vacuum Ripple Dynamics in play.

We mistook the tools of measurement for the truth of the universe. And in doing so, we built a castle of concepts on the sands of metaphor.

Closing Insight: The seduction of success is dangerous. It doesn't demand truth—it only demands utility. But when the illusion works, we stop seeing it *as* an illusion.

And thus, the myth deepens.

In the next section, we'll see how this illusion became not just a belief, but a doctrine—fueling an entire century of physics built on energy-as-entity rather than ripple-as-reality.

1.4 The Snowballing Error – The Illusion of Substance

What began as an abstract tool of utility soon mutated into a universal causal agent—as real as matter, as omnipresent as light, and as unquestioned as gravity.

The journey of this illusion didn't happen overnight. It snowballed, layer by layer, reinforced by some of the most celebrated developments in the history of physics.

Let's deconstruct how Newtonian mechanics, thermodynamics, and Einstein's relativity—each brilliant in isolation—unintentionally built a myth when combined uncritically.

Newton Gave It Motion

In Newton's world, work and force became mechanical quantities—measurable, calculable, visual. Work done = Force × Distance. Simple, elegant.

But it introduced an early confusion: if work can be measured, it must be something.

This gave birth to an idea: Energy must be real, because it moves things.

But Newton never claimed energy as substance. It was a derived quantity, an emergent property of motion—not a fundamental entity.

Thermodynamics Gave It Texture

Next came thermodynamics, which formalized energy in engines, heat flow, and entropy.

Here, the mistake deepened:

- Energy is transferred.
- Energy is conserved.
- Energy is stored.

We started treating energy like a fluid: As if it fills a system, leaks out, gets absorbed, or evaporates.

Steam engines and calorimeters made it look convincing—but still, no one had seen energy.

We had only seen heat, motion, expansion, compression—real-world phenomena. But instead of analyzing the field behaviors, we simply added up energy as a ledger balance, and assumed that ledger was the cause.

Einstein Gave It Substance

Then came the final stroke: $E = mc^2$ — elegant, powerful, misunderstood. This equation blurred the line between energy and matter. Suddenly: Energy could become mass. Mass could become energy. A subtle but dangerous message spread: Energy is a substance after all. If it can turn into matter, then surely it's *something*?

But this was never Einstein's claim. He only mapped a proportional relationship under relativistic motion—not a physical conversion of one entity into another.

Yet public understanding shifted.

Now, energy wasn't just a quantity—it was the universal stuff behind everything. Unseen, untouchable, but everywhere. Like a ghost that powers the universe.

Modern Consequences – The Myth Persists

This snowballing illusion of energy as substance has infected every level of modern thought:

- ***Climate Science Missteps:***
  - We speak of "energy imbalance" in Earth's atmosphere as if energy were a fluid being "trapped" or "leaking out."
  - But in truth, these are ripple-field misalignments, material responses to field tensions, not fluid leaks.
- ***Technology Myths:***
  - Advertisements boast "high energy storage," "clean energy transfer," and "energy-efficient design."
  - But what's being stored or transferred? Not energy as substance—but organized ripple interactions, electrical configurations, chemical tensions.
- ***AI and Consciousness:***
  - We even say "the brain uses energy like a machine," or

"consciousness is an emergent property of energy flow."
- But in UED, the brain is a ripple processor in a field matrix—not a combustion engine. The metaphor is archaic.

UED's Clarification – Energy Is the Name, Not the Thing

From the UED perspective, this entire snowball is built on a false center.

Energy is not a substance. Energy is not what moves. Energy is not stored, transferred, or released. Energy is a label—assigned to specific patterns of ripple tension, alignment, and resolution.

All physical change is ripple interaction. What we called "energy" is merely our limited projection of deeper field adjustments.

We measured the effect, gave it a name, and mistook that name for the cause.

Closing Thought:

What if the "substance" of energy never existed? What if we built a tower of technology, philosophy, and physics on a misinterpreted shadow?

The illusion continues because it works just enough to blind us. But if we wish to rethink energy, we must first dismantle this false object—this imagined "thing" that never was.

In the next chapter, we will revisit how Einstein's brilliance unintentionally cemented this myth—and how modern physics still dances around a concept it cannot define.

1.5 The Einstein Enchantment – When an Equation Became a Religion

It began as a brilliant insight—a mathematical expression that elegantly related mass and energy: $E = mc^2$.

What Einstein proposed in the early 20th century as a consequence of relativistic mechanics would soon become the most recognized formula in all of science. Simple, seductive, and powerful, it claimed that mass could be converted into energy and vice versa. This wasn't just physics; it was revelation. And the world embraced it with near-religious fervor.

But herein lies the danger. When a symbol becomes scripture, critique is heresy.

The equation worked in bombs, in stars, in nuclear reactors. That was enough for science and society to canonize it as fundamental truth. We stopped asking what it *meant*, and focused only on what it *did*. We assumed its success confirmed its ultimate reality. But under deeper scrutiny—especially through the lens of Universal Energy Dynamics (UED)—this becomes problematic.

The Illusion Deepens:

- We began to think of energy as a thing.
- We forgot the context of the equation: a mathematical correlation within a specific field theory, not a universal identity.
- We ignored the fact that we had no idea what energy truly *was*.

Energy became a currency—a mysterious, invisible substance that could be extracted, stored, transferred, and transformed. Einstein's equation was no longer a relational map between variables—it became the Genesis verse of modern physics. "In the beginning, there was energy..."

But under UED, this enchantment begins to dissolve.

UED's View:

- There is no standalone "mass" or "energy" as entities.
- What we perceive as energy release (say, in nuclear fission) is the *resolution of ripple tension* within the Structured Vacuum Energy (SVE) field.

- Mass is not compacted energy—it is *a ripple configuration held under tension.*
- Energy isn't flowing—it's *ripple stress being realigned.*

In this view, $E = mc^2$ is not wrong, but it's incomplete and misinterpreted. It tells you the *amount* of ripple tension that will be released when a bound configuration is dissolved—but it says *nothing* about the cause, structure, or mechanism of that tension. It's like quoting a fuel gauge without understanding combustion.

The equation is useful, yes—but only if we remember what it is: an approximation, not a deity. The enchantment must break for science to evolve.

1.6 Retrospective – What We Could Have Seen

Let's step back. Not from the equations or machines—but from the assumptions they fed.

What if, instead of inventing the concept of energy as a mysterious transferable "substance," early thinkers had asked a different question?

What if Clausius, Joule, Kelvin, and Einstein had paused—right before defining energy—and said:

"What exactly is moving here? What is trying to settle? What are we *measuring* when we say energy is released?"

What if, instead of naming the result "energy," they had noticed a more primal truth— that motion, temperature, light, and change were all visible signatures of imbalance seeking restoration?

That what we observe isn't the travel of some invisible stuff...but the *ripple of alignment*—a field trying to calm itself down?

The Road Not Taken

By adopting "energy" as the answer, we closed the question. We shifted from phenomenon to formalism. From *why* something happens to *how much*.

And the price we paid?

- We replaced understanding with accounting.
- We built theories that balance ledgers but blur causality.
- We began describing tension as "stored energy" rather than exploring what's actually stored: ripple configurations in a strained field.
- We confused transfer with resolution.
- We started teaching equations instead of teaching the field.

Entire industries followed. Engineers, scientists, teachers, and policy makers all inherited this framing. Children grew up believing energy "flows," "moves," "powers," "exists."

But no one has ever seen energy. What we've always seen are effects—heat, light, motion, fatigue, collapse.

UED reframes all of it—not by denying the math, but by restoring the reality behind it.

The Subtle Challenge

This chapter began with a birth. A concept. A placeholder. And now, it ends with a challenge:

"You've never seen energy. You've only seen tension resolve."

If that feels uncomfortable, good.

Because the journey ahead is not about comfort. It's about undoing illusions woven into the very language of science.

Energy was a myth. The ripple is not.

Chapter 2: Einstein's Equation – Mass Hysteria

When Mathematics Became Magic, and the World Believed the Trick.

It's just five characters: **E = mc²**. Yet these five symbols caused a psychological landslide that buried generations of thinkers under a beautifully simple myth. This equation didn't just reshape science—it rewired culture. Politicians, artists, spiritual gurus, and Silicon Valley CEOs alike invoked it as if it were divine scripture.

But here's the unsettling truth: almost no one understands what it *actually* means. Not the teacher writing it on the blackboard. Not the Nobel laureate referencing it in a speech. And certainly not the general public that reveres it like a mystical law of nature.

We turned it into a creed: **mass is energy**, energy is real, and science has cracked the code of existence. But what if this "mass-energy equivalence" is just a clever mathematical exchange rate—*not a deep physical truth*? What if the energy never existed to begin with?

What Einstein whispered as a thought experiment, the world screamed as gospel. Welcome to the hysteria.

2.1 The Most Famous Equation on Earth

If there's one equation the world knows by heart, it's **E = mc²**. It's the only bit of physics most people can recall with confidence, even if they have no idea what it means. From graffiti walls in Paris to the backs of science fair T-shirts, this five-character formula has become a global symbol for intelligence, genius, and the mysterious power of the universe.

From Patent Office to Pop Culture

The equation originated not in the frenzy of war or under the spotlight of global attention, but quietly—almost as a footnote—in a 1905 paper by a 26-year-old clerk at the Swiss Patent Office. Albert Einstein, through his special theory of relativity, proposed that mass and energy were not fundamentally different things, but convertible. One could be turned into the other, given the right conditions, and the rate of that conversion was governed by the speed of light squared: *Energy = mass × (speed of light)*^{2}.

At first, it was a curiosity. But then came nuclear fission, Hiroshima, Nagasaki—and suddenly, that elegant formula was no longer academic. It had exploded into reality, both literally and culturally.

Overnight, $\mathbf{E = mc^2}$ transformed from theoretical abstraction into *icon*. It became the cornerstone of nuclear physics, the warning on reactor doors, the badge of brilliance. The media romanticized it, poets quoted it, and governments invoked it. The equation became a symbol of the 20th century's Faustian bargain: knowledge so powerful, it could end the world.

Why the Equation Was So Seductive

At a mathematical level, $\mathbf{E = mc^2}$ has a kind of aesthetic perfection. The symmetry is irresistible: energy (E), mass (m), and the speed of light (c)—three fundamental concepts—tied together with a clean, deceptively simple relation. It's unit-consistent. It works across scales. And it made predictions that aligned with observed phenomena, like mass defects in atomic nuclei.

But in this elegance lay a trap: we mistook mathematical form for physical truth.

The Equation Doesn't Say What You Think It Says

Contrary to popular belief, $\mathbf{E = mc^2}$ does not tell us that "mass is energy" in a literal, physical sense. It's not saying that mass *is made of* energy or vice versa. Instead, the equation acts like a conversion rate—a currency exchange between two accounting systems in physics. Just as one might say "1 USD = 83 INR," the equation tells us how much energy is equivalent to a certain mass if that mass is entirely converted (e.g., in a nuclear reaction).

That's it.

It doesn't explain *how* or *why* mass transforms. It doesn't point to any underlying mechanism. It's a mathematical equivalence, not a causal law. Yet science textbooks, popular media, and even physicists often treat it as a magical axiom of nature—a gateway to understanding reality itself.

And So the Hysteria Began...

By elevating $\mathbf{E = mc^2}$ from useful equation to ontological truth, physics walked into a philosophical fog. It began to treat mass and energy as *things*—substances that could be stored, transferred, released—without ever defining what they physically *are*. The ripple tension that drives reality was ignored. Instead, we gave mass and energy a throne, crowned them with math, and called it enlightenment.

This chapter begins the necessary undoing. Let us not just question the equation—but how we've used it to build a mythology out of numbers.

2.2 What Einstein Really Meant

The world made Einstein a prophet of energy, but few ever read his scriptures carefully. The myth of $E = mc^2$ has grown so large that it often obscures what Einstein was actually trying to say. Let's rewind and return to the original context—the real soil in which this equation grew.

Context Is Not Optional: Special Relativity and Inertial Frames

Einstein's 1905 paper titled *"Does the Inertia of a Body Depend Upon Its Energy Content?"* was not a manifesto about atomic bombs or universal energy soup. It was a logical extension of special relativity, built on the idea that the laws of physics should look the same in all inertial frames—coordinate systems moving at constant speeds relative to one another.

At its heart, special relativity was a cleanup job. It redefined how space and time behave when objects move at high velocities. Within that framework, Einstein noticed something curious: if an object emits energy (say, in the form of light), its mass must decrease slightly to conserve momentum. This wasn't metaphysics—it was bookkeeping.

The now-famous equation came as a subtle consequence, tucked quietly into his appendix. It was derived under very specific assumptions:

- The system must be isolated.
- The mass must be at rest in its own frame.
- The energy is not "moving" but exists as a mathematical value associated with rest.

In short, $E = mc^2$ wasn't a general declaration about reality—it was a conclusion about symmetry and inertia under certain well-defined conditions.

Mass as "Condensed Inertia"

Einstein never said mass is "made of" energy. That's a modern exaggeration. Instead, he showed that mass behaves as if it contains energy, or more precisely, that mass carries an energetic *potential* due to how it resists acceleration—inertia.

You could say mass is inertia condensed, stabilized within a frame. It doesn't act like a fuel tank that stores energy and pours it out at will. It's not something you can scoop, bottle, or weaponize on its own.

Yet over time, this nuanced interpretation was discarded. People began speaking as if mass could explode into energy, like some magical substance. But Einstein was pointing to a mathematical consequence, not a physical transformation mechanism.

E Is Not a Substance. It's a Value

Energy in this context is not a glowing thing or a sparkly fluid. It's a numerical quantity representing the potential to do work under certain physical conditions. That's all.

The equation tells us that if you were to convert the entire rest mass of an object into usable energy, this is how much you'd theoretically get. But the equation doesn't explain *how* such a conversion happens—or whether it even can in practice, outside of very special field configurations like nuclear decay or annihilation events.

In fact, in most stable systems, mass doesn't convert into energy. It stays what it is—locked in structure, aligned in equilibrium, simply *existing*.

The Misinterpretation Was Ours, Not His

Einstein did not create the hype. He laid out a beautiful, consistent relation that worked within the narrow domain of relativistic mechanics. But the world interpreted it through the lens of wonder and war.

We took a mathematical tool and turned it into a cosmic statement: "Everything is energy." "Mass is just frozen light." "Atoms are mini power plants."

These are not Einstein's words. They are cultural poetry.

The truth? Einstein's equation measures—it does not explain. It relates mass to its energy-equivalent under idealized conditions, but it does not reveal the actual mechanism of mass, nature of energy, or cause of transformation.

That's where UED steps in—to define what he left undefined. Not to oppose Einstein, but to continue what he carefully began but never fully concluded.

2.3 The Myth of Mass–Energy Equivalence

The phrase "mass is a form of energy" has become a cornerstone of popular science—repeated in classrooms, documentaries, and even government policy papers. But what if we told you that this statement, taken literally, is false? That it's not physics but a metaphor stretched beyond its domain?

It's time to debunk one of the most persistent misconceptions in science: the idea that mass and energy are interchangeable in a literal, physical sense.

Where It All Went Wrong

Einstein's equation, $E = mc^2$, was meant to convey a mathematical relationship under specific assumptions—not to equate mass and energy as *things*. But as the 20th century unfolded—fueled by atomic explosions, scientific sensationalism, and the rise of "Big Physics"—the distinction collapsed in public discourse.

Physicists, too, fell into the trap. The phrase "mass-energy equivalence" began to circulate, suggesting that mass *is* energy in disguise, and that one can be transformed into the other like ice into water.

But that interpretation is deeply flawed. It's not just a simplification—it's a category error.

"Mass Turns Into Energy" Is a Metaphor, Not a Mechanism

In particle annihilation or nuclear decay, we often say "mass turns into energy." But what really happens is a change in configuration—a shift in how fields, particles, or forces are arranged in space. The net energy before and after remains conserved, but the mass values change because the system becomes less bound or more aligned.

Mass is not fuel that explodes into an energetic mist. It is a measure of resistance to acceleration within a field—the inertia of a system in its rest frame. When systems shift to a new configuration (such as when a nucleus splits), that difference in inertia manifests as emitted energy.

So when people say "mass becomes energy," they're using figurative shorthand for something far more subtle: a rearrangement of ripple structures that releases energy, not because mass has disappeared, but because the alignment state has changed.

Why Calling Mass "a Form of Energy" Is a Mistake

Mass and energy are not "forms" of each other like ice and water. They are quantities that happen to be related by a conversion factor (c^2), much like Fahrenheit and Celsius are related by a formula—but nobody claims Fahrenheit *becomes* Celsius.

Just because two quantities share units (joules, in this case) doesn't mean they are the same physical reality.

Here's the key point:

Energy is a scalar measure of the system's dynamic potential. Mass is a structural characteristic of how that system holds alignment in stillness.

One is a function of configuration; the other is a function of inertia. They relate—but they don't equate.

Shared Units, Not Shared Identity

The units of mass (kilograms) and energy (joules) become interchangeable in the equation $E = mc^2$ because c^2 has units that bridge them. But this is a mathematical bridge, not an ontological one.

Just like:

- 0°C = 32°F doesn't mean Celsius *is* Fahrenheit,
- 1 kg × c^2 = a certain number of joules doesn't mean mass is energy.

This confusion happens when we let unit conversion imply substance transformation. That's bad physics—and even worse philosophy.

What UED Clarifies

Under Universal Energy Dynamics (UED), energy is not some stored fuel inside mass—it is the result of ripple tension and alignment within the Structured Vacuum Energy (SVE) field. Mass is the net coherence of these ripples in a localized pattern. When a system becomes less bound, the ripple misalignment reduces, and the released field response appears as "energy emission."

So yes, the numbers on both sides of the equation match. But no, mass and energy are not two states of the same "thing." They are measurable outcomes of different field conditions.

2.4 Ripple Tension – The Actual Source of Nuclear Energy

In mainstream physics, nuclear reactions—whether fission or fusion—are explained through the language of *mass being converted into energy*. This interpretation stems directly from a superficial reading of Einstein's famous equation, $E = mc^2$, which has been widely interpreted as a license to believe that matter can be destroyed to produce energy.

But what actually happens inside a nuclear reaction?

The Hidden Reality: Ripple Collapse, Not Substance Loss

In a nuclear fission event, a heavy nucleus (like Uranium-235) splits into smaller nuclei. The resulting products have a combined mass less than the original. This "missing mass" is dubbed the mass defect, and it's said to have been "converted into energy."

But UED argues: nothing is converted. No mass disappears. No energy flies out from some inner vault of matter. What we're witnessing is a structural ripple reconfiguration.

- Mass, under UED, is a *stabilized ripple tension* in the Structured Vacuum Energy (SVE) field.
- When a nucleus splits, the internal field configuration collapses and realigns into a lower-tension structure.
- The so-called "binding energy" is not something *stored* or *released*—it is simply a measure of how much ripple tension gets dissolved during reconfiguration.

Fusion and Fission: Two Sides of the Same Ripple Game

In fusion, small nuclei combine to form a larger one, such as hydrogen fusing into helium in the Sun. Again, the resulting mass is lower than the inputs. From the standard view, mass has been "converted into energy."

Under UED, this again signals a ripple resolution:

- The incoming particles are individual ripple structures.
- When they combine, certain tensions cancel due to constructive alignment.
- The net reduction in stabilized ripple load appears as released radiation—but there's no mass becoming energy.
- This radiation is just the residual adjustment ripple—the excess field ripple needing to dissipate after the new balance is achieved.

Binding Energy: A False Phrase

The term *binding energy* is itself misleading. It implies that something physical is "stored" and later "released." But what if binding energy is just the cost of ripple alignment?

- You didn't "bind" anything—you simply reached a more stable equilibrium.
- You didn't "release" anything—you witnessed field reconfiguration in a visible way (radiation, heat, etc).

Binding energy is just the mathematical consequence of reduced ripple strain. No substance moves. No energy flows like fluid. It's a realignment event, not a transportation event.

UED's Sharp Correction:

Let's be clear. According to UED:

- Mass is not "stored energy." It is a ripple pattern stabilized under specific field constraints.
- Nuclear reactions do not destroy or convert matter. They restructure ripple configurations, changing the field's tension topology.
- The "energy" we observe (e.g., light, heat, gamma rays) is the visible effect of ripple readjustment, not a magical appearance of "converted energy."

So Why Did We Believe Otherwise?

Because energy as a concept was too convenient—and because the results (explosions, heat, power) matched the math. We never asked what actually caused those effects. We were too busy balancing the books of conservation.

Key Takeaway: *Nuclear reactions do not convert mass into energy. They collapse ripple tensions into new equilibria. What you see as "released energy" is just the universe snapping back into alignment—one ripple at a time.*

2.5 Cultural Hysteria and Scientific Drift

From Equation to Icon

In the early 20th century, Einstein's $E = mc^2$ was more than just a physics formula. It became a cultural lightning rod, a symbolic equation that transcended science and leapt into the collective imagination of humanity.

But what followed was not progress—it was hysteria.

As the equation seeped into public consciousness, it stopped being a hypothesis to be refined and became a superstition, almost a scientific religion:

- "Mass is energy!"
- "We can get infinite power from a speck of matter!"
- "Einstein explained the universe!"

This wasn't science. It was blind cultural enchantment.

"Unlimited Energy from Mass!" – The False Promise

Popular science began promoting the idea that mass—solid, physical, tangible matter—was some kind of vault of infinite power. That if only we could crack it open, the universe would pour energy into our hands.

This delusion gave birth to:

- Atomic bombs—worshipped as proof of the equation.
- Nuclear energy—sold as the ultimate future of power.
- Science fiction fantasies—fueling dreams of warp drives, teleportation, antimatter reactors.

And behind all of it, one magical incantation: $E = mc^2$.

Quantum Mysticism and Pseudoscientific Parasites

As the cultural obsession grew, pseudoscience crept in.

From "thoughts create energy" to "manifest your reality through frequency," a parade of misinterpretations hijacked Einstein's prestige. "Energy" became a spiritual buzzword. "Mass" became metaphor.

The equation was no longer understood—it was invoked.

It mutated into everything from:

- New Age healing ("your cells vibrate with energy!"),
- to conspiracy theories ("free energy is being suppressed!"),

- to motivational slogans ("You are pure energy!").

Einstein didn't predict this. But culture did. Because when science loses philosophical grounding, it becomes myth in a lab coat.

The Real Cost: Scientific Drift and Complacency

This mass-cultural glorification of $E = mc^2$ created a dangerous side effect: A deep intellectual laziness.

- Why explore deeper causes when you already have "the ultimate truth"?
- Why question the nature of mass or energy if the equation explains it all?
- Why challenge Einstein when he's become the untouchable prophet?

This intellectual comfort zone calcified scientific thinking. Instead of evolving new models, we began worshipping a model.

Energy became a substance. Mass became a source. And equations became beliefs.

The ripple-based reality—subtle, dynamic, structural—was ignored. And the Structured Vacuum Energy—the real stage of all interaction—was forgotten.

The Comfort of Closure

Human psychology craves certainty. And $E = mc^2$ gave it in one line.

It promised:

- Simplicity in a complex world.
- Predictability in a chaotic universe.
- Control over the forces of nature.

But in doing so, it also offered closure—and closure is the death of curiosity.

It seduced science into believing it had reached the mountaintop. When in reality, it had only set up camp at the base of the real terrain.

UED's Correction: Energy is not stored in mass. Mass is a ripple configuration in a structured field. No substance is converted—only alignment shifts.

The universe is not made of "things that become other things." It is made of ripples seeking balance.

Closing Reflection: *"We replaced religious myth with mathematical myth—then called it enlightenment."*

—Sandeep Chavan

2.6 Consequences of the Illusion

When a myth enters the bloodstream of civilization, it doesn't stay in the lab. It leaks into classrooms, boardrooms, war rooms—and minds.

Einstein's $E = mc^2$, misunderstood and mystified, has become more than a misused formula. It is now a cultural virus that continues to distort our thinking about the nature of matter, mind, machines—and meaning.

Let's decode the consequences.

Misguided Metaphors: When Science Meets Fantasy

The misuse of $E = mc^2$ birthed a new language of illusion. Metaphors meant to explain became misconceptions adopted as truth:

- **"The body is energy."** No—it's a collection of structured ripple configurations working in biological harmony. Calling it "energy" without context is lazy reductionism.
- **"Thoughts have mass."** False. Thoughts are ripple alignments in neural fields, not matter. But once you treat

energy and mass as interchangeable fairy dust, any poetic phrase gains the illusion of scientific backing.

- **"AI with energy potential."** The new tech hysteria. Machines do not become 'energetic beings' just because they compute. Confusing logic flow with energy flow is the modern echo of Einstein worship gone wrong.

These aren't just sloppy analogies—they are conceptual drift that erodes precision and opens the door to pseudoscience.

Education: Where We Teach the Myth

From the earliest school levels to elite universities, $E = mc^2$ is presented as gospel. But how is it really taught?

- **Oversimplification:** Textbooks compress complex interactions into slogans. "Mass becomes energy" is repeated without ever discussing what mass truly is or how energy behaves at the ripple level.
- **Romanticization:** Students are told: *"This is the most beautiful equation in physics."* Beauty is fine—but not when it replaces questioning with admiration.
- **Memorization, not Realization:** Rather than explore the foundations of the field, students are trained to chant equations like mantras. The equation becomes a ritual. UED is absent from the syllabus.

What is left behind is not an educated mind, but a conditioned one.

Policy: The High Cost of Wrong Assumptions

Science myths don't stop at culture or classrooms—they write laws, shape military strategy, and guide billion-dollar decisions. The illusion of mass-energy conversion has infected global policy:

- **Nuclear Strategy:** Political leaders and strategists often speak of nuclear weapons as *"releasing the stored energy of matter."* This framing is not only conceptually flawed—it fosters false confidence in control over consequences.
- **Energy Policy:** Governments invest in reactors, fusion fantasies, and quantum technologies based on the dream of unlimited energy from matter. But if energy isn't a substance in the first place, the entire architecture of expectation is flawed.
- **Climate Models:** When models are built on assumptions where energy is "emitted" or "absorbed" as a quantifiable substance, rather than ripple field changes through ATT and alignment, then we are not predicting reality—we are approximating illusions.

The ripple is ignored. The structure of vacuum is misunderstood. And policies ripple those misunderstandings forward.

The Deepest Cost: Intellectual Blindfolding

The worship of $E = mc^2$ created a condition worse than ignorance: An illusion of knowledge.

- We stopped asking *what is energy?*
- We never questioned *what is mass made of?*
- We never wondered *how does a field align to cause form, behavior, and perception?*

Because we thought we already knew.

The equation was supposed to open doors. Instead, it closed inquiry, sealed by the cement of admiration.

UED's Correction: Mass is not converted into energy. Nothing is destroyed, and nothing is created. The universe reconfigures ripple alignments under the law of equilibrium. ATT

(Alignment-Time-Tension) governs interaction—not mass-energy voodoo.

We don't extract power from mass—we realign fields to release tension within structures.

Closing Thought: *"The greatest illusion of modern science was not what it didn't know—but what it thought it knew." —Sandeep Chavan*

2.7 Retrospective – Beyond the Equation

Let us pause and look back—not at the man, but at the myth his equation became.

For over a century, $E = mc^2$ has stood like a cathedral in the cathedral of physics: grand, symmetrical, and seemingly eternal. But what if this monument, for all its beauty, has obscured something more fundamental? What if we mistook the stained glass for sunlight?

Rewriting the Premise

What if we had never thought of mass as a "form of energy"? What if we had interpreted Einstein's insight not as a conversion from one physical substance to another—but as a transition in field states?

In this alternate lens, mass isn't something that "becomes" energy. Mass is a configuration—a ripple tension held in place, like a rope pulled taut. It's not fuel. It's form. When nuclear events occur, what's released is not some mystical substance called energy. What's happening is a realignment—a collapse of ripple tension that allows equilibrium to restore itself in a new configuration.

Ripple Tension vs. Energy

We've inherited the language of "energy release," but this is a relic of industrial metaphors. Steam engines release energy. Coal releases energy. But what if atoms don't release anything? What if the so-called "energy" is merely a reduction in stored ripple tension, redistributed through the structured vacuum as new field configurations?

By framing mass as *frozen ripple alignment* rather than an energy container, we open a door to cleaner, causally sound models—without the magic.

Unlearning the Spell

We must be willing to say: The equation was elegant. The physics was revolutionary. But the interpretation... was romanticized.

The consequence? A century of false equivalence, where the mystical idea of "mass turning into energy" captured imaginations more than it answered questions.

Ending Thought

"Mass is not energy waiting to be released. It is tension waiting to resolve—if the universe allows."

That resolution isn't ours to command. It belongs to the field. And when we finally listen to it, we may stop chasing myths and start hearing the music of the ripples beneath.

Chapter 3: Quantum Confusion

"Light is a wave!" "No, it's a particle!" "Wait... it's both!"

That's not physics. That's intellectual panic disguised as profundity.

For over a hundred years, quantum mechanics has asked us to hold contradictions in our head like holy relics. Not to resolve them. Not to examine their cause. But to believe in them—because the math works.

The infamous double-slit experiment became the altar of this confusion. Shine light through two slits, and you get an interference pattern—like a wave. Place a detector to see "which slit" the light went through, and suddenly the interference vanishes—like a particle. So what is light?

A wave? A particle? Or some magical entity that shapeshifts depending on your curiosity?

Instead of re-examining our model of reality, we forced a marriage between wave and particle—a shotgun wedding officiated by desperation. Thus was born the doctrine of wave-particle duality. Not a solution, but a slogan.

And then came the photon—a hypothetical "packet of light," discreet, quantized, and supposedly bouncing through space like a cosmic tennis ball. A beautiful concept for equations. A disaster for understanding.

This wasn't clarity. This was linguistic camouflage for deeper ignorance.

Quantum theory, as it stands today, is not wrong in its numbers. It predicts extremely well. But it describes horrendously. It does not explain light. It circumvents light with clever tricks, renaming the gaps as mysteries to be tolerated.

In mainstream physics, this has become the standard escape clause:

"Quantum mechanics is weird. Don't try to understand it—just calculate."

That is not science. That is surrender.

We won't fix this confusion by adding more quantum jargon, nor by glorifying indeterminacy as a feature. We will not add another patch to the tattered quilt.

We are going to dissolve the confusion at its root.

By peeling off the illusion layer by layer—starting with the false ontology of the photon, and ending with a new perspective based on Structured Vacuum Ripple Alignment—we will expose the truth quantum theory tried to hint at but never saw.

There is no particle. There is no duality. There is no collapse. There is only ripple tension. And its attempt to resolve.

Let's begin.

3.1 The Myth of the Photon Packet

It began with heat. Or more precisely, with the way objects glow when heated.

In the early 20th century, physicists encountered a troubling anomaly. When they studied the radiation emitted by hot objects—like molten metal or stars—they couldn't explain why the emitted light behaved the way it did. Classical physics predicted a catastrophic infinity of energy at short wavelengths—a disaster known as the ultraviolet catastrophe.

Enter Max Planck.

In an act of mathematical desperation, Planck proposed that light wasn't emitted continuously, but in discrete amounts—quanta. These "chunks" of energy, he suggested, were proportional to the frequency of the radiation. He didn't know what these quanta really were. He didn't even believe in them physically. To Planck, it was just a trick to fix the math.

But then Einstein took the trick seriously.

In 1905, Einstein applied this quantized idea to explain the photoelectric effect—a phenomenon where light hitting a metal surface ejects electrons. Classical wave theory couldn't explain why the energy transfer was instant and threshold-dependent. Einstein proposed a radical solution:

"Light behaves like it's made of particles. Discrete packets. Quanta. Let's call them... photons."

And with that, the photon was born—not as an observed entity, but as a mathematical workaround. A conceptual Band-Aid turned into an entire belief system.

The Particle of Energy? A Dangerous Misstep

Over the next century, the photon was promoted from metaphor to entity. Students were taught that light was made of "particles of energy"—self-contained, massless, zipping through space like bullets made of brightness.

This did two things:

1. It gave physics a neat image—something to imagine, manipulate, and teach.
2. It also broke everything.

Because the moment you treat light as a particle, you're forced to explain:

- Why it diffracts.

- Why it interferes.
- Why it doesn't have a definite path.
- Why you never see "half a photon."
- Why photons can appear to arrive without crossing any space (quantum tunneling).
- Why no one has ever seen a photon in isolation.

The more you look, the less sense the "particle" metaphor makes.

And yet, the photon survived—propped up by math, not understanding. We began assigning energies to photons, polarizations, spin states, even trajectories. But beneath the surface, it was still just a label for observed effects, not a physical object.

Like a ghost that knocks over a chair—we saw the result and gave it a name. But never the thing itself.

UED Correction: No Packet, No Bullet, No Bounce

In Universal Energy Dynamics (UED), light is not a packet of energy. It is not a particle. It is not a "thing" moving through space.

It is a ripple. Not a wave of something—but a tension alignment propagating through the Structured Vacuum Energy (SVE) field.

When an atom "emits light," it's not launching a photon like a cannonball. It's undergoing a realignment—a sudden drop in local ripple tension. That drop disturbs the field, and the disturbance ripples outward like tension in a stretched fabric.

There is no flying object. There is only the field, responding to imbalance.

And when another system "absorbs" this ripple, it is not catching a photon—it is completing the alignment. The ripple reaches it, the system shifts, and equilibrium is updated.

The entire "emission–travel–absorption" model is just a chain of ripple responses. Like tuning forks across a room vibrating in sympathy—not because something hit them, but because the field between them carried tension.

Analogy: Lighting a Lamp Without Throwing Water

Imagine this:

You strike the surface of a still pond with a stick. Ripples spread outward.

Now, think of trying to light a distant lamp by throwing a water droplet at it. Sounds absurd, right?

That's what the photon model wants you to believe—that light travels like thrown particles to hit and illuminate something.

But in reality, you don't need to throw anything.

You touch the field—you disturb the medium—and the ripple carries the effect.

Light is the ripple. The field is the medium. The ripple does not carry energy inside it like a container. It is the energy—as alignment tension, moving across a structured fabric.

And once you see this, the photon vanishes—not in failure, but in clarity.

Closing Note for This Section

The photon was never a thing. It was a metaphor born of mathematical emergency.

We mistook a ripple for a rock—and then spent a century trying to make the rock behave like a ripple.

UED invites us to dissolve the packet illusion and listen to the field itself.

The light doesn't travel. It resonates.

3.2 Wave–Particle Duality – A Forced Marriage

"Light is both a wave and a particle." "It depends on how you observe it." "Quantum mechanics defies classical logic."

These statements are not explanations. They are concessions.

A century ago, physicists were faced with a dilemma. Some experiments showed light diffracting, interfering, bending—clear evidence of wave behavior. Others, like the photoelectric effect or Compton scattering, hinted at discrete energy interactions—as if light behaved like a particle.

So what did physics do?

It didn't solve the contradiction. It didn't revise the model. It invented a compromise—a linguistic loophole:

Wave–Particle Duality. The logic went like this:

- If you see an interference pattern → it's a wave.
- If you detect a discrete impact → it's a particle.
- And if you see both effects in the same experiment → it's both!

Instead of re-evaluating what light truly is, physicists did what desperate systems always do when faced with irreconcilable views: they married them. Not because it made sense—but because it kept the math working and postponed the philosophical reckoning.

The Patchwork That Became Doctrine

Wave–particle duality is not a discovery. It's a patch—a conceptual Band-Aid over two incompatible interpretations.

Here's how the patch was stitched together:

1. Wave-like behavior (e.g., double-slit interference) was undeniable.
2. Particle-like outcomes (e.g., discrete electron ejection) were observable.
3. The math for both worked, but each model broke down in the other's domain.
4. Rather than unify the cause, physics declared:

"Light is sometimes a wave and sometimes a particle—depending on how you ask the question."

It was clever. It was flexible. And it was utterly incomplete.

Because what kind of entity morphs its nature based on your experiment? What kind of thing is and isn't a particle at the same time?

Instead of finding a third explanation, we institutionalized the contradiction. And worse, we taught it as truth.

UED Response: The Duality Is an Illusion

Universal Energy Dynamics (UED) dissolves this false marriage at its root.

There is no duality.

There is only ripple behavior in a structured field, and the illusion of particlehood arises from how the observer interacts with the ripple—specifically, through ATT thresholds (Alignment–Time–Tension).

Let's decode it:

- What we call wave behavior is simply field-wide ripple alignment—coherent tension propagating across the Structured Vacuum Energy (SVE).
- What we call particle behavior is the localized spike of interaction—a sharp change in alignment at the observer's interface.

But the key insight is this:

The "particle" never travels. It doesn't exist in transit. It only appears at the point of detection, where the field tension meets the observer's ATT limit.

The so-called particle is not a thing. It's a consequence. A measurement artifact. A localized result of ripple collapse under field tension.

Just like a lightning bolt is not a "thing" moving through air—it's a sudden breakdown of insulation in the medium—what we see as a photon "hitting" a detector is simply the alignment threshold breaking down, forcing the system to readjust sharply.

That adjustment is what we *record* as a particle.

But nothing ever traveled.

The Problem Was Never Light. It Was Language.

Wave–particle duality is a linguistic illusion—a framing error.

- We framed light in terms of objects: "Is it a thing?"
- We framed behavior in terms of classical categories: "Does it act like this or that?"
- We tried to reconcile measurement spikes with continuous field behaviors, using metaphors instead of mechanisms.

UED invites us to reframe it all:

- Light isn't a wave.
- Light isn't a particle.
- Light is a ripple alignment response in a medium that doesn't move, but transmits through field strain.

The duality disappears when we stop projecting object-based assumptions onto ripple-based behavior.

Final Analogy: The Shadow Puppet Illusion

Imagine watching a shadow puppet show. You see shapes—animals, people, objects—moving across the screen.

But behind the curtain, there's only the dancer's hand and a light source. The illusion is created by interaction—not by the puppets being real entities.

Wave–particle duality is a shadow puppet. A surface illusion projected by our own apparatus, our language, and our obsession with substance.

When we look behind the curtain, we don't find a photon toggling between identities.

We find field tension responding to misalignment.

That's not duality. That's ripple intelligence in action.

3.3 What Actually Collapses – Not a Wavefunction, but Your Expectation

"Before measurement, the particle exists in all possible states. After measurement, the wavefunction collapses into one outcome."

So goes the standard quantum tale.

It's a story told with conviction—and deep unease. Because unlike other physical theories, quantum mechanics demands an observer not just to record reality, but to create it. Measurement is no longer passive; it is fundamental. It causes the collapse.

But what collapses, exactly?

Physicists tell us it's the wavefunction—a mathematical expression encoding all the possible outcomes of a quantum system. But this "collapse" is never defined as a physical process. It has no speed, no duration, no mechanism. It's just assumed:

At the moment of measurement, all possibilities vanish, and one result "materializes."

Magic.

Not a theory. A ritual.

We never see the wavefunction collapse. We only see a detector click, a particle trace, a data point appear. From this, we assume that something collapsed somewhere—and that this event defines reality.

This is not physics. This is narrative.

The Collapse Was Never Physical – It Was Epistemic

The wavefunction is not a real wave. It is a tool—a probabilistic map of possibilities based on what we know. When we observe a quantum system, the "collapse" does not occur in the system—it occurs in our knowledge of the system.

What really collapses?

Not a particle. Not a wave. Not space or energy.

What collapses is uncertainty—our inability to predict, finally replaced with an observed value.

But this collapse is cognitive, not causal. And yet, quantum mechanics treats it as a physical event without a physical explanation.

UED View: Collapse Is Alignment Completion

In Universal Energy Dynamics (UED), we throw out the idea of wavefunction collapse altogether.

Here's what's really happening:

- Every system is a ripple configuration in the Structured Vacuum Energy (SVE) field.
- Every observer is a localized ripple field with its own ATT threshold—a specific limit of Alignment–Time–Tension beyond which it cannot maintain coherence.
- When a ripple interaction reaches an observer, measurement happens when the field tension crosses that ATT threshold.
- This isn't collapse. It's completion—a field interaction realigning itself in a new configuration.

So the moment of "collapse" is not the system losing possibilities—it is the field achieving resolution within the observer's alignment structure.

No Possibilities Died. No Realities Collapsed.

There were never "many states" hovering in limbo. That's a mental model projected from statistical math onto real ontology.

There was only field imbalance propagating through the vacuum.

What we perceived as indeterminacy was not cosmic randomness. It was observer misfit—our inability to resolve field interactions until the ripple finally aligned with our ATT limits.

Let's reframe it:

- A ripple travels across space, modulating field tension.
- The observer (e.g., a detector) has a built-in threshold of how much misalignment it can tolerate.
- When the ripple crosses this threshold, the system realigns sharply.
- That snap-back is what we record as an "event."

It didn't collapse reality. It resolved field tension.

And we, the observers, mistaken that for the universe making a choice.

Quantum Indeterminacy Is Observer-Relative Misfit

The infamous unpredictability of quantum events—the randomness in photon arrival, electron spin, particle decay—is not evidence of a chaotic universe.

It is a symptom of mismatched alignment between:

- The ripple tension in the field.
- The configuration capacity of the observer.

The more precisely your ATT field is tuned to the incoming ripple, the more consistent the outcome.

The more misaligned or constrained it is, the more "random" the result appears.

What we call indeterminacy is really a blind spot—a local ripple mismatch, not a universal law.

Analogy: Tuning a Radio

Imagine tuning an analog radio. As you turn the dial, you hear static, then partial voices, then suddenly a clear station locks in.

That "snap" from static to clarity is not the station collapsing into existence—it was always broadcasting. What changed was your receiver's alignment.

Quantum measurement is the same.

The field was always rippling. Your detection was the moment you finally matched the ATT configuration to resolve a coherent interaction.

Collapse? No. Completion? Yes.

Final Thought: The Collapse Was Always Yours

The wavefunction didn't collapse. The universe didn't split. The photon didn't decide.

You did.

Your instrument, your threshold, your perception forced a choice—not on the cosmos, but on the interaction. What followed was a resolution event, encoded in ripple logic.

The mystery dissolves when you stop imagining particles in boxes, and start seeing the field as a dynamic canvas of tension seeking balance.

What collapses isn't the wave. It's the illusion that there ever was one.

3.4 Emission and Absorption Reinterpreted

In the mainstream model, this is the story we're told:

- An atom gets "excited."
- One of its electrons jumps to a higher energy level.
- Later, the electron drops back down—and in doing so, it "releases a photon."
- That photon, now zipping through space, may eventually collide with another atom.
- Upon collision, the second atom "absorbs the photon," causing one of its electrons to jump up.

A tidy, transactional tale—almost like Newtonian billiards in the quantum world. Energy in. Energy out. One object throws, another catches.

It feels intuitive. Visual. Simple. But it's not reality.

The Problem With the Mainstream View

This photon-packet narrative, though widespread, suffers from deep conceptual flaws:

- Where is the photon between emission and absorption? No one has ever detected a photon mid-flight without collapsing it.
- What is the photon? A particle? A wave? A hybrid? No one can agree.
- How does a point-like entity carry frequency, polarization, direction, and energy—all at once—with no internal structure?
- What medium does it travel through? Space? Vacuum? Some undefined field?

To resolve these, physicists began layering metaphors:

- "It's a wave-packet."
- "It behaves like a probability cloud."
- "It's a quantum excitation of the electromagnetic field."

And yet, no one can show the photon. Only its detection.

In other words: the photon doesn't exist between events. It exists only at the interaction points—before and after. Everything in between is abstract, hypothetical, and interpretative.

UED Reframe: No Packet, No Bullet—Only Ripple Realignment

Universal Energy Dynamics (UED) dismantles this illusion completely.

There is no "emission" in the sense of launching a particle. There is no "absorption" in the sense of catching one.

Here's what's actually happening:

Step 1: Field Tension Builds

An atom enters an unstable ripple configuration—call it "excited," but what it really means is: ripple tension has increased in its localized zone of the Structured Vacuum Energy (SVE) field.

Step 2: Realignment Triggered

The atom can't hold that tension forever. At some threshold (governed by its ATT: Alignment–Time–Tension condition), it snaps back to a lower-tension state. This is not a "release"—it's a collapse of misalignment.

Step 3: Ripple Propagation

That collapse doesn't throw a particle. It disturbs the field. A ripple emerges—spreading as a tension wave through the SVE. Not as a thing traveling through space, but as a field readjustment moving outward.

Step 4: Receiver Field Matches

Elsewhere, another atom exists in a configuration that is receptive to this exact ripple frequency and geometry. When the incoming ripple resonates with its ATT structure, a lock-in happens.

That atom "absorbs" the ripple—not by grabbing a photon, but by completing the alignment initiated by the first atom.

This is not transfer. It is ripple handshake. A field-wide reconfiguration. An entangled dialogue—not a transaction.

No Particle Exchange – Just Ripple Continuity

Under UED, emission and absorption are not isolated events but two ends of a ripple response.

The ripple is continuous, field-aligned, and invisible unless you disturb it. Like tension traveling across a tight rope—you don't see the rope move, you feel the ripple at the other end.

There's no photon flying through space. There's no handoff of a tiny energy ball. There's just alignment restoration—the universe adjusting itself to relieve tension.

Why This Solves Long-Standing Mysteries

Coherence

In mainstream optics, coherence refers to the phase relationship between emitted photons. But what keeps a particle in phase with others? How can discrete bullets sync?

Under UED, coherence is field geometry continuity. Since ripples travel through the same vacuum structure, their phase coherence is a result of shared origin and path—not mysterious synchronization.

Polarization

Photons are said to have spin and polarization, but how does a point-particle exhibit directional vector properties?

In UED, polarization emerges naturally as the orientation of ripple tension in the vacuum field. It's not assigned—it's encoded in the ripple's geometric structure.

Directionality

Mainstream models struggle with how a photon "knows" where to go or how to remain focused in a laser beam.

In UED, the ripple propagates along existing field pathways of minimum resistance—like stress waves through a crystalline lattice. The direction is not chosen; it is shaped by the field's geometry and boundary conditions.

The Real Picture: Emission Is You Relaxing. Absorption Is Me Feeling It.

Here's a more human metaphor.

Imagine you're tense—emotionally charged. You take a deep breath and exhale. That release doesn't fly across the room like a bullet.

But someone in tune with you—someone emotionally connected—feels it. They absorb your state change. They realign, without needing to touch you.

That's light in UED. That's the real emission and absorption.

Final Reflection: The Universe Doesn't Throw—It Resonates

The transactional model of quantum light—of photon bullets—was a projection of our mechanical mind onto an intelligent field.

But the universe doesn't shoot. It doesn't throw. It doesn't exchange particles.

It resolves. It aligns. It ripples.

What we call "light" is not a substance being passed from one atom to another. It's a shared memory of tension, encoded in the vacuum, looking for resolution.

And when that alignment completes—it glows.

3.5 The Measurement Illusion

Quantum mechanics made one radical move that forever changed how we talk about reality:

It put the observer at the center of physics.

No longer was the universe unfolding independently. Now, the act of observation became entangled with the outcome. A particle had no position until measured. A system remained in superposition until disturbed. The observer became the final authority—the cosmic referee who decides what becomes real.

But this raised a question that quantum theory never truly answered:

Who or what is the observer?

Is it a human being? A machine? A camera? A field? A conscious mind?

Mainstream physics doesn't say. It offers only vague responses:

- "Anything that causes irreversible interaction."
- "A macroscopic system that amplifies the result."
- "A detector that triggers decoherence."

But all of these are just functional descriptions—not ontological ones. None define what an observer is in terms of physics itself. The term is left hanging—a ghost in the theory, collapsing wavefunctions without explanation.

UED Correction: The Observer Is a Ripple Field With ATT Limits

Universal Energy Dynamics (UED) fixes this foundational gap.

The observer is not a person, a sensor, or a machine. It is a localized ripple field, embedded within the Structured Vacuum Energy (SVE), defined by its ATT threshold:

- Alignment
- Time
- Tension

These are not metaphysical categories—they are physical limits.

Let's break it down:

Alignment

Every system in the universe has a unique ripple pattern—its "signature geometry." This determines what kinds of external ripple configurations it can resonate with or respond to.

Time

Every system has a temporal coherence zone—a time window during which it can maintain an unresolved tension before readjustment becomes inevitable.

Tension

There is a threshold of ripple imbalance that the system can tolerate. Once this threshold is crossed, the field must realign, creating a measurable effect.

Together, these three factors define the measurement boundary—the point at which a ripple interaction becomes visible, recorded, or felt by the system.

What You "See" Is Not the Universe Changing—It's Your Field Resolving

In standard quantum mechanics, a measurement is treated as an event that collapses the system. But this is an illusion created by:

- Our misunderstanding of what is interacting.
- Our insistence on viewing the universe as made of particles.
- Our reliance on detectors rather than field logic.

In UED, what we call measurement is not something happening "out there." It is a local resolution event inside the observer's ripple zone.

Here's how it works:

1. A ripple disturbance (light, field, particle event) propagates through the SVE.
2. The observer field is exposed to this disturbance.
3. If the ripple stays below the observer's ATT threshold, nothing happens. It passes through silently—unfelt, unrecorded.
4. If the ripple matches or exceeds the threshold, the field can no longer hold its equilibrium.
5. The observer realigns—this internal collapse is what we perceive as detection, observation, or measurement.

But nothing "collapses" in the universe. Nothing is chosen or forced into a state. Nothing jumps from superposition into certainty.

Your field resolves. That's all.

Measurement Is a Consequence, Not a Cause

Mainstream physics often implies that the observer causes change in the system—forcing it into a particular outcome.

UED inverts this:

The system is already rippling. The observer's field is the one being pushed into resolution. The "result" is a record of the observer's limits, not the universe's nature.

This is why different observers can see different things under different conditions. Reality doesn't contradict itself. Observers interpret ripple differently because their ATT thresholds vary.

Analogy: The Tensioned Drum Skin

Imagine the universe as a vast drumhead—silent, still, but under latent tension. Every atom, every system, every observer is a circle of skin stretched over part of this drum.

Now, strike one part of the drum. A ripple propagates.

If your skin is loosely stretched (low ATT), the ripple might pass without effect. If it's taut (high ATT), it will vibrate violently when the ripple hits. That vibration is what you call a measurement. But the ripple was there either way.

You didn't collapse the wave. You simply responded to it.

The "event" happened because your field reached its limit, not because the universe had to pick a reality for you.

UED's Elegant Resolution of the Observer Paradox

Under UED, there is no measurement problem.

There is only tension resolving through alignment—with thresholds defined by structure, time, and resonance.

The observer is no longer mystical. No longer divine. No longer needed to "complete" the universe.

You are not the creator of the result. You are the last node in the ripple chain, finally snapping into place.

Final Thought: You Didn't Witness the Universe Collapsing—You Felt It Realign

The quantum illusion was never in the wavefunction. It was in the assumption that observation is a one-way gaze into reality.

In truth, measurement is a two-way handshake—between ripple and receiver.

And when the handshake completes, you say, "I saw something." But what really happened was this: "The ripple fit you—and you flinched."

3.6 The Real Problem: Language, Not Light

Physics, at its best, is an attempt to describe the universe with clarity and precision. But somewhere along the way, quantum mechanics forgot the first rule of clarity:

Words matter.

In trying to describe things we couldn't see, physicists began using language that was metaphorical, vague, and sometimes outright misleading. These weren't just harmless figures of speech. They became mental traps. They created an entire generation of scientists, teachers, and students who spoke fluently in paradoxes—without realizing they were built on fiction.

The Vocabulary of Confusion

Let's examine some of the most common phrases in quantum mechanics:

"Quantum Jump"

Sounds exciting—like an electron vaulting from one orbit to another. But this phrase is entirely metaphorical. There is no literal "jump." No visible trajectory. No intermediate motion. Just an abstract change in energy state, wrapped in a dramatic verb.

Yet, the phrase stuck. It made electrons sound like acrobats. And now it's part of our cultural vocabulary—even used in business presentations and self-help books.

"Photon Hit"

As if light were a tiny bullet, slamming into an electron or a metal plate. Again, no one has ever seen a photon mid-flight. What we detect is a response, a tension release, a field interaction. But by calling it a "hit," we turn ripple interaction into an object-based collision.

It's language borrowed from billiard balls, not ripple fields.

"Probability Cloud"

This term is used to describe the region where an electron is "likely to be." But it evokes an image of a fuzzy fog, as if the electron were smearing itself out over space. In reality, this is just a statistical tool—a map of likelihoods, not a physical cloud. The "cloud" has no texture, no edges, no substance.

But the term made it sound mystical, soft, and mysterious.

And so, we built an entire mythos around vague labels.

Quantum Was Never Weird—Our Interpretation Was Lazy

The real tragedy of quantum physics is not that it's mysterious, but that it was prematurely declared mystical.

Why?

Because the math worked, and the metaphors were convenient.

Rather than investigate the structure of the field, or question whether "particles" actually existed, physicists leaned on probabilistic abstractions and linguistic shortcuts. They gave names to gaps in understanding—and then worshipped the names.

So "duality" became an answer. "Collapse" became an event. "Uncertainty" became a law. This wasn't bold science. It was intellectual abdication.

The Ripple Reality Hides Behind the Words

Universal Energy Dynamics (UED) shows that once you strip away these metaphors and look directly at the structure beneath, the weirdness fades.

What we called a "quantum jump" is actually: A ripple configuration reaching a tension threshold and reconfiguring.

What we called a "photon hit" is: A ripple handshake between two field structures across the SVE.

What we called a "probability cloud" is: A map of possible ripple alignments—not the spread of a ghostly object.

There is no mystery—only mislabeling. There is no magic—only language pretending to be reality.

Why This Matters: Language Becomes Law

When you describe something long enough in the wrong terms, those terms start shaping your worldview. In physics, these terms have become so entrenched that they dictate what questions are allowed, what models are built, and what kind of answers feel acceptable.

Because of this:

- New researchers inherit broken assumptions.
- Students are taught paradoxes instead of principles.
- Critics of the standard model are silenced—not by logic, but by jargon.

The language has become a gatekeeper—not to protect truth, but to preserve an illusion.

UED's Correction: Describe Behavior, Not Belief

UED proposes a simple shift:

Stop describing light and matter as things. Start describing how the field behaves.

This is not poetry. This is precision.

- Don't say "a photon arrived." Say: "Ripple tension propagated and triggered a realignment."
- Don't say "the electron jumped." Say: "A field zone crossed its ATT limit and reconfigured."
- Don't say "we collapsed the wavefunction." Say: "Our field resolved its misalignment."

This language may sound new at first. But it's honest. It doesn't pretend to know more than we do. It doesn't insert invisible particles just to satisfy a metaphor.

Final Reflection: The Problem Was Never the Physics—It Was the Phrasing

We didn't misunderstand the universe because it was too strange. We misunderstood it because we named it poorly.

Once you remove the fictional packet, the imaginary jump, the mystical cloud—and replace them with ripple logic and field alignment—the contradictions dissolve.

Light is no longer weird. It's just misunderstood. And physics becomes, once again, not a myth... but a map.

3.7 Closing Reflection – A Light Without Particles

"If you stop believing in photons, the light doesn't disappear— it just stops pretending to be something it's not."

Physics, for all its complexity, has one sacred responsibility: to tell the truth about reality, not just in numbers, but in understanding.

For over a century, light has been forced to play a dual role it never auditioned for. It was cast as both wave and particle, as both smooth and bumpy, as both flowing and bouncing. But all this time, light was not confused—we were.

We mistook our models for the thing itself. We mistook our labels for laws. And we mistook ripples in the field for flying beads of energy.

But here's the truth, revealed by Universal Energy Dynamics (UED):

Light never "traveled." It was never a photon zooming through space. It was never a bullet. Never a blob. Never a package.

It was, and always has been, a ripple alignment. A tension response in the structured vacuum. A consequence—not of something moving—but of something trying to settle.

Stop Patching. Start Listening.

Quantum mechanics tried to reconcile contradiction with cleverness. It invented photons to solve equations. Then it invented duality to explain the failure of photons to behave.

When that wasn't enough, it created wavefunction collapse, decoherence, quantum jumps, hidden variables, many worlds—and still, light kept breaking the narrative.

It's time to stop fixing the old map.

UED doesn't offer another patch. It offers a new lens. It says:

- There is no particle.
- There is no packet.
- There is no arrival.

There is only field tension realigning—and your detection of that realignment.

You don't need to believe in photons. You need to feel the ripple.

The Light That Never Left

Light was never sent. It was never emitted like a gift-wrapped energy box tossed into the void.

What really happened?

One configuration changed, disturbing the field. Another configuration resonated, completing the change.

In between, nothing flew. But everything aligned.

This is not romantic minimalism. This is ontological clarity.

When You Let Go of the Particle, the Pattern Becomes Clear

The universe doesn't use packages. It doesn't postmark energy from atom to atom. It adjusts, resonates, and responds through structured ripple logic.

Light is not a thing. It is a gesture. A field memory, reawakening at the right moment.

And when it aligns with your field—your ATT threshold—you call it "seeing."

But light never came.

You and the universe aligned.

Final Word: Toward a Particleless Worldview

Imagine a world where light needs no messenger. Where detection is not interception, but interaction. Where reality is not made of pieces, but of patterns shifting through a living field.

That world is not imagined. It is here. You were just taught to ignore it.

Let's restore our vision. Not with better math, but with deeper honesty. Because the moment we stop insisting that light is a particle, we don't lose anything.

We gain the universe itself—unwrapped, unfiltered, and finally understood.

Interlude – Leaving the Illusion Behind

We have spent the first part of this journey dismantling a world built on metaphor.

We questioned energy as a substance. We unraveled mass as a ripple configuration—not a vault of hidden power. We re-examined light—not as a particle or wave, but as ripple tension seeking resolution. We showed that quantum confusion was not a feature of the universe, but a feature of language, assumption, and unresolved models.

It wasn't the universe that was weird. It was our insistence on fitting the universe into mechanical, object-based logic. We called it energy. We called it particle. We called it wavefunction.

And in doing so, we built a magnificent tower—on air.

What we inherited was not a map of reality, but a script of convenience. It helped us build machines, bombs, theories, and industries. But it never told us what energy really is.

Because energy was never a thing. It was a consequence.

A tension. A restlessness. A ripple in the stillness of something far more fundamental.

And that is where we now turn.

In Part II, we shift from demolition to discovery.

We step beyond the metaphor, and into the medium itself. Not to describe the world through approximation, but to feel how it functions—at the most intimate level of ripple alignment and field tension.

We'll explore:

- Why energy is not stored—but strained.
- Why atoms don't release power—but restore balance.
- Why nothing flows—but everything resonates.
- And why stillness, not movement, is the true foundation of dynamic reality.

This is no longer about revising physics. It's about remembering the field.

Welcome to the real universe.

Welcome to UED.

Part II: UED – Energy Redefined

"Nothing is energy until it is resolved as alignment tension at the observer's field."

Chapter 4: The Real Nature of Energy

"You never had energy. You only had imbalance."

For centuries, we spoke of energy as if it were the ultimate substance—the invisible fuel behind all movement, all change, all power. We tried to capture it, store it, transfer it, and convert it. But no one ever held a jar of energy. No one ever found it hiding behind a proton or leaking out of a broken machine.

What we saw—what we always saw—was tension. We saw fields under stress. We saw systems trying to realign. And we gave that effort a name: energy.

But energy is not something that is. It is something that happens.

This chapter marks a turning point. No more metaphors. No more approximations.

We are no longer patching holes in old theories. We are reframing reality.

Under **Universal Energy Dynamics (UED)**, energy is not a substance. It is not even a property. It is a ripple response—a field-level disturbance that only appears when the system strays from its equilibrium.

In this new lens, energy does not exist on its own. It occurs—as the ache of stillness trying to return.

We're about to define energy not as a quantity to be conserved, but as a symptom of field imbalance. A precise, measurable, yet entirely emergent event within the Structured Vacuum Energy (SVE) fabric.

And once you see this, you'll realize: Energy was never the source. It was always the signal.

Let's begin.

4.1 The Mistake of Trying to "Find" Energy

For over two centuries, scientists have acted like treasure hunters—digging, probing, smashing particles, wiring detectors—hoping to "find" energy as if it were some hidden object tucked inside the universe.

We were told energy is *stored* in fuels. That it *resides* in atoms. That it *flows* through circuits. That it's *transferred* from one system to another.

And so, naturally, we began looking for it—as if it were a thing.

But no one ever found energy. What we found were effects: heat, light, motion, voltage, deformation, collapse. We measured what changed—and inferred that something *caused* the change.

So we gave that invisible cause a name: energy. And from that naming, the hunt began.

The Illusion of a Substance That's Never Seen

Ask a child, "Where is energy stored?" and they'll point to a battery. Ask a physicist, and they'll explain "energy levels" or "potential wells" or "mass-energy content." Ask an engineer, and they'll talk about energy efficiency, losses, and transfers.

But probe deeper—what is the thing being stored, transferred, or lost?

There is no answer. Just a swirl of language:

- "Kinetic energy" – motion.
- "Potential energy" – position.
- "Thermal energy" – vibration.

- "Nuclear energy" – binding configuration.

None of these are substances. They are states. Descriptions of tension. Labels on systems not at rest.

So what have we been storing, transferring, and converting?

Nothing.

We've only been disturbing alignment—and watching systems attempt to restore it.

From Storage to Strain: The UED Reframe

In Universal Energy Dynamics (UED), there is no "stored energy."

There is only stored tension—a ripple configuration that is temporarily misaligned within the Structured Vacuum Energy (SVE) field.

A battery does not contain energy. It contains chemical imbalance—a ripple asymmetry that will push electrons when released.

A moving object doesn't carry kinetic energy. It carries directional ripple strain—a vector of field misalignment.

A hot object isn't filled with thermal energy. It's filled with micro-ripples struggling to equilibrate.

This is the critical transition:

From imagining energy as a static thing to understanding it as a dynamic struggle.

We do not "possess" energy. We momentarily inhabit a misaligned state.

And when systems move, glow, or heat—it's not because energy flowed. It's because alignment was disturbed and is trying to return.

Why We Got It Wrong

The classical worldview was rooted in material metaphors:

- Mass as substance.
- Force as push.

- Energy as fuel.

This made sense in the age of steam engines and coal. You burned fuel, wheels turned, and the work was done. The concept of energy worked—*as a bookkeeping tool.* It helped us predict, measure, optimize.

But the more precise our observations became—at atomic, quantum, and field levels—the more this model failed to explain what was actually happening.

We began seeing things that couldn't be flowing, yet still caused effects. We measured energy without any visible substance. We observed systems "exchanging" energy without anything changing hands.

The old metaphor had broken down.

And instead of discarding it, we tried to stretch it—to patch it with photons, virtual particles, statistical clouds, and field quantizations.

UED says: enough.

The metaphor didn't break because reality is strange. It broke because energy was never "there" to begin with.

Final Insight

We've spent generations trying to locate energy. But energy has no location. It has no shape, no container, no independent existence.

It is not something you can find—because it only appears when something breaks equilibrium.

You cannot store it. You cannot isolate it. You cannot move it from one place to another like cargo.

What we called "stored energy" was always just stressed geometry in a ripple field. What we called "released energy" was the unwinding of that stress.

And what we felt—what we named "power"—was simply the relief of tension finally allowed to realign.

We were never hunting for a thing. We were witnessing a process. And giving it the wrong name.

4.2 Energy Is Ripple Tension

Let's speak plainly: Energy is not a thing. It is not fluid. It is not mass. It is not stuff.

It is ripple tension—a measure of how far a system has been pulled away from its natural alignment with the field.

That's it.

In the language of Universal Energy Dynamics (UED), energy is simply the field's discomfort—a quantifiable deviation from equilibrium in the Structured Vacuum Energy (SVE) field.

Whenever you say "something has energy," what you are really saying is:

"It is strained." "It is unresolved." "It is out of tune."

No Substance, No Packet, No Flow

Mainstream science has long described energy as if it were a hidden currency moving through invisible pipes:

- "Electricity flows through wires."
- "The sun sends energy to Earth."
- "We store energy in fuel."

But under UED, none of these metaphors hold. Because there is no flow. There is no packet being tossed across space. There is only field reconfiguration.

Think of energy not as water in a hose, but as tension in a net.

When one part of the net is disturbed—plucked, twisted, compressed—that tension doesn't "move" in the traditional sense. It redistributes, attempting to restore balance across the whole structure.

That redistribution is what we perceive as:

- Light.
- Heat.
- Motion.
- Vibration.
- Decay.

All are just different expressions of the same thing: ripple tension trying to dissolve.

Tension Is the True Signature of Energy

Let's reframe the classic categories of energy using UED:

Kinetic Energy

Classically: the energy of motion.

UED: directional ripple tension within a moving configuration, trying to stabilize through interaction.

Potential Energy

Classically: energy stored by position or configuration.

UED: built-up ripple strain in a displaced field geometry—waiting for realignment.

Thermal Energy

Classically: energy due to random particle motion.

UED: chaotic ripple disturbance in microfield structures—localized tension radiating through imbalance.

Chemical/Nuclear Energy

Classically: energy stored in bonds.

UED: interlocked ripple geometries—stabilized tensions that, when broken or merged, release alignment imbalances.

All these are not different "forms" of a universal fluid called energy. They are different textures of ripple tension, shaped by the systems involved.

Why This View Works

Because it's consistent.

- It does not require imaginary carriers like photons or

gluons.
- It explains why energy always correlates with *change*—because change is what happens when tension resolves.
- It removes the paradox of energy appearing and disappearing in "virtual" processes—those are just field strains that didn't reach the ATT threshold to register as events.

UED doesn't abstract energy. It grounds it—directly in the behavior of the field.

Once you recognize that the vacuum itself is structured, with geometric integrity, the idea of energy becomes obvious:

Any deviation from that structure will produce strain. And that strain—when measured—is what we used to call energy.

But we don't need the name anymore. We just need the pattern.

You Only See Energy When It's Unhappy

A system perfectly aligned with the SVE doesn't glow, heat up, or move.

It is calm. Still. Silent. Balanced.

But disturb it, and that stillness ripples. That ripple is what we detect:

- A spark of light.
- A jolt of current.
- A bang.
- A shift.
- A wave.

Every observed effect is just this:

The field trying to relax.

Final Reflection

We chased energy across atoms, stars, and accelerators. But what we were really chasing was tension—the field's subtle rebellion against imbalance.

What we mistook for power was pain. What we stored was strain. What we released was realignment.

Energy is not what a system has. It's what a system feels when it drifts from where it belongs.

Once you stop trying to catch energy, and start tracking ripple tension, the universe becomes simpler—more honest.

Because energy is not the fuel of reality.

It is the footprint of its friction.

4.3 The Equation of Occurrence: UED's Master Relation

Physics has always tried to capture reality through equations—symbols arranged to reflect the underlying order of the universe. But most classical equations describe energy as if it were an object: something to be measured, stored, converted, and conserved.

They tell us:

- $E = \frac{1}{2}mv^2$ – kinetic energy.
- $E = mgh$ – gravitational potential.
- $E = mc^2$ – mass-energy equivalence.

These equations are functional. They work—for machines, for rockets, for bombs.

But they do not explain what energy is. They only assign numbers to outcomes of interaction.

None of them tell you:

- Why energy arises.
- Where it comes from.

- What causes it to appear or disappear.

They do not describe energy as an event. They describe it as a quantity—as if energy were some universal currency waiting to be spent.

Universal Energy Dynamics (UED) changes that.

Energy Is Not a Number. It Is a Ripple State.

In UED, energy is not a stored variable. It is the field's response to misalignment. It is not what the system *has*. It is what the system *experiences* when it cannot rest. And this experience can be modeled with a simple yet profound relation:

$E \propto \Delta T \times A$

Where:

- E is the ripple energy—the measurable tension present in the system.
- ΔT is the duration of the misalignment—the time the field remains out of tune.
- A is the amplitude of misalignment—how far the ripple deviates from its stable geometry.

This is the UED Master Equation—the fundamental way to understand energy not as a fixed unit, but as a consequence of field tension over time.

Let's unpack this.

ΔT – The Time Component: Duration of Discomfort

A ripple that disturbs the vacuum for a brief moment may never register. The field might restore itself before any significant effect occurs.

But if the misalignment persists, if the field is held out of equilibrium for a duration, tension begins to build.

That's ΔT: the temporal stretch of imbalance.

Energy doesn't "exist" at a moment. It accumulates over time as the system stays strained.

This is why short, intense pulses feel different from slow, steady disturbances—because time shapes tension.

A – The Amplitude: Degree of Misalignment

Not all ripple disturbances are equal.

A small field nudge may cause only a tiny adjustment. A large deviation—such as a nuclear misalignment or electric imbalance—creates immense tension.

This is A: the field deviation—how far the system strays from its aligned, relaxed configuration.

The bigger the misfit, the more violent the re-alignment—and the more energy you observe.

This is why a tightly packed nucleus or compressed spring appears "high in energy"—it is not holding a substance, but sitting in a state of unnatural geometry, full of ripple stress.

Putting It Together: $E \propto \Delta T \times A$

This equation says:

The experienced energy in any ripple system is proportional to:

- How far it is out of balance (A)
- How long it remains that way (ΔT)

This is not a currency to be spent. It is not a resource to be transferred. It is a field state trying to resolve itself.

Energy appears when ripple tension accumulates. And it vanishes when realignment completes.

It doesn't move. It doesn't jump. It occurs—in the same way a stretched rubber band doesn't "have energy," but simply aches to return to its shape.

Why This Is Not Just Another Formula

Most classical energy equations are context-dependent:

- Kinetic energy only applies to mass in motion.
- Potential energy only applies in a gravitational or elastic field.
- Heat equations depend on temperature and mass.

But the UED equation is universal.

It doesn't depend on the object. It depends on the field tension surrounding it.

You can apply it to:

- A vibrating molecule.
- A black hole.
- A battery.
- A wire.
- A brain.

Because in all cases, what you're measuring is ripple tension held over time.

And once released, that tension appears as light, sound, motion, collapse, or reconfiguration—not because something moved, but because something realigned.

Final Reflection

The UED Master Equation is more than a new energy formula. It is a redefinition of what energy is.

Energy is not what you use. It's what the universe feels when it's pulled out of harmony.

$E \propto \Delta T \times A$

A ripple stretched for too long, too far—will snap.

And when it does, that snap is what we used to call "energy release."

But now we know: It was just equilibrium restoring itself—beautifully, precisely, and without ever sending a single particle.

4.4 ATT Logic: Alignment–Time–Tension

If energy is not a thing, not a flow, not a substance—then what is it really?

In Universal Energy Dynamics (UED), energy is understood through the logic of ATT:

Alignment – Time – Tension

This is not just a clever acronym. It is the trinity of conditions that must exist for energy to *appear*.

Let's be precise:

- Energy does not "exist" in stillness.
- Energy is not always present.
- Energy is not fundamental to the universe.

It is conditional. It is emergent. It is felt only when something has drifted from its natural rest.

ATT explains when and why that feeling arises.

A — Alignment (or Misalignment)

All systems, from subatomic particles to galaxies, exist within the Structured Vacuum Energy (SVE) field. This field has natural configurations—preferred geometries where ripples hold coherent balance.

When a system is in alignment, there is no tension. No glow. No motion. No force.

But the moment a ripple structure deviates—bent, compressed, displaced, twisted—it falls out of sync with the surrounding field.

This is misalignment.

It's like tuning a guitar string just a bit too tight or pressing a wrinkle into an otherwise smooth fabric. The field *remembers* its harmony—and now, it wants to fix the break.

That deviation from equilibrium is the first seed of energy.

Without misalignment, there is nothing to resolve—no tension, no dynamics, no "event."

T — Time (ΔT) — Duration of Disturbance

Misalignment is not enough. To manifest tension, it must persist.

If a ripple disturbance arises and immediately resolves, there's no measurable energy. It's like a single raindrop hitting still water and vanishing.

But if the misalignment endures, if the system holds its deviation—due to confinement, force, boundary, or inertia—the ripple begins to strain the field.

The longer it resists returning to alignment, the more tension accumulates.

This is ΔT—the duration of discomfort. The time component in the UED Master Equation.

Time is not just a coordinate here. It is the amplifier of field imbalance.

Just as a held breath creates pressure, a held misalignment builds potential.

T — Tension — The Ripple Strain That We Mistook for Energy

Finally, when alignment is broken and the field holds that misfit over time, what builds up is tension—not metaphorical, but physical.

Ripple tension is the stored misalignment encoded in the local structure of the SVE.

This tension can be:

- Local (like in a compressed spring).
- Distributed (like thermal agitation in a gas).
- Relational (like gravitational distortion across cosmic scales).

But in all cases, what we experience as energy—as heat, motion, voltage, vibration—is just the release of this ripple tension.

That's why energy always "does work." Because it is not passive. It is the unraveling of imbalance into resolution.

Tension is not a side effect of energy. Tension is energy—as it truly exists.

ATT: The Full Picture

Putting it together:

1. Alignment gives us the *geometry* of equilibrium.
2. Time gives us the *duration* of misalignment.
3. Tension gives us the *felt strain* of that condition.

Together, they form a field condition, not a particle state. And only when all three are present and exceed a certain threshold, does energy appear.

This is why some disturbances remain invisible, while others release light, sound, or heat. The difference lies in the ATT threshold being crossed.

No threshold crossed → no observable energy. Threshold exceeded → measurable event.

Rethinking Energy as Field Memory

Under UED, energy is not just a number—it is memory + strain:

- A record of misalignment.
- A trace of disturbance.
- A tension held too long, too far.

It is not substance, but symptom. It is not a particle, but a story—encoded in the language of field dynamics.

Once you see this, the entire classical vocabulary collapses:

- There's no such thing as energy "contained" in an object.

- No meaningful "conversion" between energy forms.
- No "losses," only redistribution of ripple stress.

Final Reflection

We were taught that energy is fundamental.

But that's like saying fever is fundamental to biology. Fever is not a substance. It's a signal of imbalance.

Energy is the field's fever—a visible reaction to hidden misalignment.

When the field forgets how to rest, it begins to ache. That ache is tension. And that tension is what we call energy.

ATT gives us the structure to understand this—not as mysticism, but as field intelligence under stress.

Once you master ATT logic, you no longer chase energy. You trace alignment.

Because the universe does not run on power. It runs on its desire to return to balance.

4.5 Why Energy "Occurs," Not "Exists"

We've been speaking about energy for so long as a *thing*—as a possession, a resource, a substance—that it's difficult to unlearn the illusion. But unlearn it we must.

Because energy does not exist in the way we were taught.

It is not a constituent of matter. It is not "inside" atoms or batteries or waves. It is not waiting to be unlocked or released like treasure from a vault.

Energy occurs—it emerges temporarily, conditionally, *only when the field is out of tune with itself.*

This distinction—between existence and occurrence—is not just philosophical. It is foundational.

Lightning Is Not Stored

Imagine a thunderstorm. Tension builds between the clouds and the Earth. Charge accumulates across a gap. And then, suddenly—lightning.

But what exactly is the lightning?

Is it a substance? A fluid? A stored object suddenly escaping?

No. Lightning is not a thing. It is an event—the visible collapse of field imbalance through a highly conductive path.

There was no "lightning stuff" hiding in the cloud. There was only electrical tension, held in place until it occurred—as a sudden, brilliant realignment.

Energy is just like that.

What we call energy is not what's stored in the cloud. It is the moment of release, the pattern of reconfiguration, the result of instability resolving itself.

We see the light. We hear the thunder. But we never touch energy—we only witness its signature.

Waves in the Vacuum—Not Objects in a Box

This same principle applies to every so-called "energy source" in science:

- A battery has no energy *inside it*—only chemical imbalance, ready to ripple when connected.
- An excited atom is not a container of light—it is a strained ripple configuration waiting for relief.
- A wound spring isn't storing energy—it is misaligned geometry under stress.

In all these examples, nothing "exists" in isolation.

There is only a field under tension—and that tension becomes visible through interaction, through release, through change.

Energy Is the Signature of Imbalance

Think of a puddle after rain. The water doesn't "contain" the wind that disturbed it. But the ripples on its surface tell you that something passed through.

In the same way, energy is not what's inside the system. It's what the system shows you when it's struggling to return to stillness.

This is the most radical idea UED presents:

Energy is not a component of reality. It is a consequence of misalignment within the Structured Vacuum Energy (SVE) field.

No tension? No energy. Perfect alignment? No motion, no heat, no force. But when the balance is disturbed, the field tries to heal—and that healing is what we perceive as energy.

Why This Matters

In the classical view, everything has energy:

- A ball at rest has "potential energy."
- A gas molecule has "internal energy."
- A photon has "energy content."

But in UED, this thinking is replaced by something more precise:

- A ball has no energy until its position interacts with a gravitational field imbalance.
- A molecule has no internal energy—only ripple disturbance held within bounds.
- A photon doesn't have energy—it *is* a momentary ripple, not a particle.

In short, there is no substance to be transferred—only ripple tension to be resolved.

The Myth of Energy Inside the System

Textbooks often talk about "energy inside the system" or "energy entering/leaving the system."

But this framing is flawed.

There is no internal tank. No measurable quantity traveling like fuel from point A to point B.

There is only a distributed ripple network—and when you change one part, the field reconfigures to accommodate the shift.

The system doesn't "contain" energy. It temporarily inhabits misalignment—and that misalignment expresses itself through ripple tension.

The Universe Doesn't Store Energy—It Expresses Alignment

Let's be clear:

- The universe is not running on stored energy.
- It is running on structural memory and response.
- Every atom, every wave, every galaxy exists in an ever-fluctuating tension field.
- What we call energy is just the universe's way of returning to coherence.

Energy doesn't *exist* in this model. It is the echo of misalignment, playing out in form and motion.

Final Insight

We spent centuries trying to possess energy. To find it. To mine it. To own it.

But now we understand:

You cannot possess what was never a thing. You can only observe what was temporarily out of place.

Energy does not sit inside particles. It is not locked in molecules. It is not emitted, absorbed, or transferred.

It occurs when the field forgets its stillness. It vanishes when alignment is restored.

And that is the most elegant truth of all:

The most powerful force in the universe is not energy—it is equilibrium, reclaiming itself.

4.6 Reframing Motion, Heat, and Change

We now arrive at one of the most liberating consequences of UED:

If energy does not exist as a substance, then the phenomena we associate with energy must also be misunderstood.

We must reframe what we think we see.

Because if energy is ripple tension in a structured field—not a transferred fluid or flowing packet—then the everyday expressions of energy we call *motion*, *heat*, and *change* are not what they appear to be.

The universe is not filled with moving things. It is filled with reconfiguring tensions, restless alignments, and ripples adjusting to equilibrium.

Let's decode them one by one.

Motion: Not Movement, But Realignment

In classical physics, motion is defined as a change in position over time.

You throw a ball, and it "moves" from your hand to the ground. A planet "revolves" around the sun. An electron "orbits" a nucleus.

This is object-based thinking—projecting the behavior of solid particles through an abstract coordinate system called space.

But under Universal Energy Dynamics (UED), there are no solid particles moving *through* space. There is only ripple alignment—changing across the structured vacuum field (SVE).

Motion is not the travel of a thing. It is the perceived shift in ripple geometry from one field configuration to another.

You never see the ball move. You see the field reconfiguring its ripple patterns to match what we interpret as "displacement."

This explains:

- Why particles never truly "travel" in quantum models.
- Why velocity can be reframed as ripple coherence direction.

- Why the same mass can appear stationary or in motion depending on frame alignment.

In UED, the concept of movement is replaced with perceptual ripple shift—a dance of the field, not a parade of particles.

Heat: Not Particle Vibration, But Micro-Ripple Turbulence

Heat has long been described as the "random motion" of particles. In thermodynamics, a hotter system simply means faster-moving molecules bouncing around, colliding, transferring kinetic energy.

But what if there are no particles? What if energy isn't kinetic? What if nothing is actually moving?

UED answers: Heat is not vibration. It is unresolved ripple tension.

A "hot" system is one in which micro-ripples are trapped in chaotic misalignment, bouncing not because they're moving freely—but because their field geometry hasn't relaxed.

Temperature, then, is not a measure of motion. It is a measure of how misaligned a system is from field equilibrium on the micro scale.

This explains:

- Why heat always moves from hot to cold (tension seeks resolution).
- Why absolute zero is not a lack of motion, but perfect alignment.
- Why different materials "store heat" differently (they tolerate misalignment differently).

In UED, what we call thermal energy is simply localized ripple agitation—anxiety in the field.

Change: Not Motion Over Time, But Field Reconfiguration

And now we arrive at the biggest shift of all.

We think of the universe as a theater of continuous change. Things age. Decay. Evolve. Expand. Explode.

But what is actually changing?

Under UED, the answer is profound:

The universe is not moving. It is reconfiguring itself.

Time is not passing. Particles are not traveling. Space is not shifting.

What's happening is the field—the SVE—is constantly rebalancing itself through ripples of tension and moments of relief.

"Change" is not the alteration of what exists. It is the restoration of what wants to exist peacefully.

That is why:

- Atomic decay doesn't "lose" energy—it drops into alignment.
- Evolution isn't random—it is ripple compatibility improving under stress.
- Entropy isn't disorder—it is reconfiguration toward ripple harmony.

We're not watching the universe tick like a clock. We're watching it breathe—through pulses of imbalance, tension, and realignment.

What This Changes in Our Understanding

If motion is ripple shift, If heat is ripple turbulence, If change is ripple reconfiguration—

Then all dynamics of the universe are field-based, not object-based.

We stop thinking in terms of:

- Masses moving.
- Energies flowing.
- Particles colliding.

And start thinking in terms of:

- Ripple patterns forming, holding, and resolving.
- Field geometries adjusting under ATT pressure.
- Observables appearing only when misalignment crosses threshold.

Suddenly, the strangeness of quantum mechanics and the friction of classical mechanics both dissolve into a single elegant model:

A universe of stillness, disturbed into motionless change.

Final Reflection

We thought the world moved. We thought particles zoomed around like billiard balls on cosmic tables. We thought heat was vibration, and change was progress.

But in the language of UED, none of these are things. They are field responses.

Motion is not the journey of an object. It is the ripple's quest to remember its place.

Heat is not the shaking of particles. It is the field's agitation while trying to calm itself.

Change is not what happens to things. It is what the field does to restore balance.

So don't ask, "What is moving?" Ask:

"Where is the field misaligned, and how is it trying to come home?"

Introducing Chavan's Principle: Energy as Resolution

Chavan's Principle: *"Nothing is energy until it is resolved as alignment tension at the observer's field."*

This principle radically redefines how we interpret all observable phenomena. It tells us:

- Energy is not a universal substance present independently in space or objects.
- It is not traveling through space in packets or flows.
- It is not absolute—it is relational.

Energy is the field's ripple disturbance. What we observe as "energy" is simply the moment when that ripple reaches and crosses the ATT threshold of an observer—causing measurable or perceivable realignment.

Implications of Chavan's Principle

No Observer → No Event

Until a ripple is resolved at an ATT-defined boundary, it is not "energy." It is field strain—silent, invisible, unfelt.

Just like:

- Sound doesn't exist in a vacuum—not because it's absent, but because no ripple receiver is present to resolve the pressure oscillation.
- Light doesn't glow unless a ripple-aligned system completes the tension collapse.
- Heat isn't felt unless it exceeds a threshold within the sensing system.

Every System Is an Observer

Not just humans. Not just instruments. Every structure in the field is an *observer*—defined by its own ATT configuration.

A molecule, a plant, a rock, a black hole—all "observe" ripple differently. When their ATT thresholds are crossed, energy occurs for them.

Energy Is Not Objective—It Is Threshold-Based

This does not mean energy is subjective. It means it is contextual, field-local, and response-based.

Why This Changes the Core of Science

Chavan's Principle reframes the act of measurement, interaction, and even perception:

- Measurement is not detecting something that exists. It is being changed by something that occurs.
- Force is not applied—it is the memory of ripple misalignment playing out across systems.
- Observation is not passive—it is participatory field reconfiguration.

4.7 Closing Insight: Stillness Is the Source

"Energy is not a resource—it's a disturbance." "The universe does not run on power—it resolves itself through pattern."

Let us pause now—not in confusion, but in clarity.

After dismantling the myth of energy as substance, and after reinterpreting motion, heat, and change through the language of ripple tension, we arrive at the most profound realization of all:

Behind every ripple... there is stillness. Beneath every vibration... there is silence. Beneath what we call energy... there is a state that wants nothing.

The Structured Vacuum Energy (SVE)—the foundation of UED—is not a seething chaos of particles and fields. It is not dark energy or invisible fluid. It is a field of silent symmetry, infinitely aligned, completely balanced, and inherently still.

And energy? Energy is what happens when that stillness is disturbed.

The Myth of the Active Universe

In classical physics, the universe is imagined as dynamic by default—constantly expanding, vibrating, burning, colliding, exploding. It's a cosmos of ceaseless doing.

We then ask: *What fuels all this?* And we answer: *Energy.*

So we built an entire worldview around energy-as-resource:

- The sun "produces" energy.
- Muscles "consume" energy.
- Machines "convert" energy.
- Civilization "runs" on energy.

But this perspective is backward.

The universe is not busy by nature. It is still by nature—and it is tension that makes it appear active.

What we've been calling energy is not the fuel of the cosmos. It's the signal that something has gone off-balance.

The cosmos isn't doing. It's recovering.

Patterns Over Power

The true mechanics of the universe are not about force, power, or speed.

They are about pattern.

Everything that emerges—atoms, waves, bodies, galaxies—is just a local configuration of ripples, trying to stay in coherence with the field.

And when those patterns are broken—by boundary, by interaction, by misalignment—the field reacts.

That reaction is what we feel as energy.

But it is not action. It is resolution.

The universe is not a machine running on power. It is a symphony seeking harmony—each disturbance causing a series of ripple adjustments until equilibrium is once again achieved.

Stillness: The First and Final Condition

Before light, before time, before space—there was stillness.

Not emptiness. Not absence. But perfect alignment. A silent, structured field with no tension, no bias, no asymmetry.

And all that we call the universe—every force, every form, every frequency—is what happened when that stillness became disturbed.

In this sense:

- The Big Bang is not an origin of matter, but a rupture of field coherence.
- Time is not flowing, but tracking how long misalignments last.
- Space is not expanding—it's being redefined by ripple stretch.

And energy, in every case, is not what creates the universe. It is the ache of the universe returning to its original state.

Why This Changes Everything

If stillness is the source, then the goal of every process is not to accelerate, dominate, or perpetuate—but to resolve.

This flips the foundation of science, philosophy, and civilization.

We've pursued energy. We've tried to conquer it, extract it, amplify it.

But what if the wise approach is not to generate more energy, but to generate less tension?

Not to fuel more reactions—but to build more resonant configurations.

Not to burn harder—but to align deeper.

Because every energy event is a cost—not of power, but of misalignment.

And true sustainability is not technological. It is ontological—about living in resonance with the field.

Final Words

Stillness is not the absence of energy. Stillness is the precondition for energy to vanish. The universe was not born to move. It moves because it remembers what stillness felt like—and it's always trying to get back.

Everything you call energy—light, heat, power, charge—is just a field in grief, mourning its broken symmetry. And every moment of alignment—whether in atoms or thoughts—is a step toward restoration.

So stop asking where energy comes from.

Ask instead: *What disturbed the stillness?* And: *What is the field trying to heal?*

That is where the real physics begins.

Chapter 5: Dissolution = Release

"Atoms don't explode because they contain energy. They release energy because they can no longer contain tension."

We were taught that inside the atom lies unimaginable power—trapped energy waiting to be unleashed. That mass can be "converted" into energy, and that nuclear reactions are nature's way of trading matter for motion.

We built bombs, reactors, stars, and scientific careers on this metaphor.

But what if it's wrong?

Not wrong in the mathematical sense—after all, the explosions happen, and the numbers add up. But wrong in the interpretation.

What if energy isn't "inside" the atom at all?

What if the atom is not a storage vault, but a delicate ripple arrangement, barely holding itself together in the Structured Vacuum Energy (SVE) field?

What if the so-called "release of energy" is not the liberation of fuel—but the collapse of unresolved tension into a lower, calmer configuration?

Dissolution, not destruction. Realignment, not conversion. Field relief—not fuel consumption.

In this chapter, we rethink fission and fusion not as energetic miracles, but as field events—natural movements of ripple geometry toward rest.

We will show that:

- Atoms are tense field arrangements, not containers of mass or energy.
- Mass loss is not a transformation—it's a reshaping of ripple interaction.
- What we call "binding energy" is simply the cost of coherence.
- And what we call "release" is just the field exhaling.

Let's now look at what really happens when atoms break, or fuse—and why the real story is not about power, but pattern resolution.

5.1 The Atomic Illusion: Stored Power or Strained Pattern?

For over a century, the dominant narrative around nuclear energy has been seductively simple:

"Mass converts into energy."

It's the most dramatic application of $E = mc^2$—a formula that turned invisible equations into mushroom clouds and starshine. We were told that matter, if broken down correctly, could erupt into vast amounts of usable energy. That inside the atom lies unimaginable power, waiting to be unlocked.

From schoolbooks to political speeches, the idea was reinforced:

- Mass defect explained why nuclear reactions lose mass.
- Binding energy was described as the glue that holds nuclei together.
- Stored nuclear force was imagined as compressed energy hidden in matter.

In this picture, atoms are like tiny batteries—charged with potential energy just waiting to be triggered.

But UED sees this very differently.

Because the moment you let go of the idea of energy as a substance, the entire atomic metaphor begins to unravel.

What Is Mass Defect, Really?

In nuclear reactions—whether fission or fusion—the mass of the resulting products is slightly less than the mass of the original particles. This "missing mass" is interpreted as being "converted" into energy, usually in the form of heat, light, and radiation.

But here's the problem: this explanation treats mass as a thing that can be lost or transformed.

In Universal Energy Dynamics (UED), mass is not a substance. It is the ripple signature of a system's configuration within the Structured Vacuum Energy (SVE) field. When the configuration changes—when a bound system becomes more relaxed—its ripple footprint changes. That's what we observe as "mass loss."

The mass didn't disappear. The system simply reconfigured into a lower-tension pattern. And the resulting difference in ripple tension was expressed as measurable release—heat, radiation, light—not because something was "converted," but because something was resolved.

Binding Energy: A Misleading Metaphor

We're told that nuclear particles are "bound" together by strong forces, and that it takes energy to pull them apart—just like separating magnets or stretching a spring. But again, this metaphor assumes there's a substance holding things together.

In UED, what we call binding energy is just a measure of field strain required to hold a ripple structure in place. A tightly bound nucleus is not "glued"—it's twisted and interlocked in a high-tension pattern.

And when that pattern becomes unstable, or gets disrupted—whether by neutron impact or fusion conditions—it can no longer maintain that strain. The field snaps into a more relaxed state, releasing the excess ripple tension it once held.

No glue. No stored power. Just misalignment finally letting go.

The Illusion of Stored Nuclear Force

The mainstream view talks about nuclear force as if it's a pressure trapped inside matter, waiting to be tapped. This has led to the common belief that atoms are containers of potential destruction—tiny bombs in waiting.

But in UED, there is no force "inside" the atom. There is only geometry under stress—ripple alignment barely holding itself together.

An unstable nucleus isn't a threat because it contains energy. It's a threat because it's desperately misaligned—and when it breaks, the field rushes to resolve the incoherence.

The energy we witness is not what was stored in the nucleus. It's what the field expresses when a ripple configuration dissolves and rebalances.

UED Challenge: These Are Not Substances

UED reframes all classical nuclear terms:

Classical Term	UED Interpretation
Mass defect	Ripple footprint reduction
Binding energy	Ripple strain required to hold configuration
Stored nuclear force	Residual misalignment held under tension
Energy release	Field realignment and ripple collapse

This shift is crucial.

Because it tells us that nuclear energy is not a release of fuel—it is the relaxation of geometry.

We've been looking at atoms like bombs, when they're really stressed-out structures trying to heal.

Final Thought: The atom is not a container of energy. It is a pattern barely holding its breath. And when it exhales, we see heat, light, and radiation—not because something exploded—but because the field finally had permission to let go.

The real question isn't how much energy is stored in matter. It's how much tension the field is still carrying.

And that brings us to the core of nuclear reactions—not as power-generating events, but as tension-resolving patterns. Dissolution, not destruction. Alignment, not annihilation.

5.2 Internal Ripple Tension in Atomic Structures

We tend to imagine atoms as miniature solar systems—tiny, stable arrangements of particles orbiting and interacting through invisible forces. This visualization, however useful for teaching, is deeply misleading.

Under Universal Energy Dynamics (UED), the atom is not a container of particles bound by magical forces. It is a ripple configuration—a tension geometry held in place by field alignment, not substance or force.

Atoms are not solid. They are not full of "things." They are stable but strained ripple formations embedded within the Structured Vacuum Energy (SVE) field.

Let's decode this from the inside out.

Protons and Neutrons: Interlocked Ripple Bundles

What we call "particles" are not balls of matter.

- A proton is a tightly compressed ripple bundle—its apparent mass is a signature of how tense and confined its internal field structure is.
- A neutron is an even more precarious configuration—neutral on the outside but carrying intense

inner ripple stress due to its unique alignment mismatch.

In UED, these "particles" are better understood as coherent distortions in the vacuum field. Their identity is not derived from mass or charge as standalone properties, but from how their internal ripple logic interacts with the surrounding field.

The idea that quarks and gluons are bouncing around inside protons is replaced by something more elegant:

Protons and neutrons are not made of smaller particles. They are interference knots—localized ripple concentrations that appear stable because their strain is evenly distributed.

But stability doesn't mean absence of tension. It means tension held in a specific, self-sustaining pattern.

Binding Energy = Field Tension in Overlap Zones

When protons and neutrons come together in a nucleus, their ripple patterns begin to overlap.

This overlap region is what physicists refer to as binding energy. But in UED, this is not energy being added or stored—it is field strain increasing due to constrained geometry.

Think of it as a weaving of misalignments. Each nucleon adjusts its position within the larger nucleus, contributing its own ripple logic, sometimes reinforcing tension, sometimes partially cancelling it. The result is a high-strain configuration that can appear stable only because the geometry is tight enough to contain the misfit.

But misfit it remains.

The nucleus does not stay together because it's strong. It stays together because its tensions are geometrically trapped.

Just like a skyscraper held in place not by glue, but by forces pushing against each other—structured strain.

Instability = Geometries That Can No Longer Cope

When the number of protons and neutrons becomes unbalanced, or the internal configuration becomes too large (as in uranium or plutonium nuclei), the field tension becomes unmanageable.

There is only so much misalignment the field can sustain before it begins to fray.

At some point, the geometry crosses the ATT threshold:

- Alignment becomes unsustainable.
- Time can no longer hold the structure together.
- Tension reaches a breaking point.

The result? The nucleus doesn't explode from within—it dissolves. The internal ripple bundle begins to collapse into more stable, lower-tension geometries, and the field around it reconfigures.

This is the moment we call fission or radioactive decay—but what's truly happening is ripple rebalancing.

The nucleus is not destroyed. It's simply realigning to survive.

Ripple View: No Forces, Just Patterns Under Strain

In this view:

- There is no strong nuclear force holding the nucleus together.
- There is no "energy" being added or removed.
- There is no glue, no gravity, no binding substance.

There is only patterned tension—ripple geometries straining against the fabric of the field, holding position as long as coherence allows.

And when the pattern becomes untenable, the field resolves the strain in the only way it knows how:

By reshaping itself—through what we observe as "release."

Final Reflection

The nucleus is not a ticking bomb. It is a compressed breath—a ripple held under pressure.

And like all things under pressure, it eventually exhales.

So when we speak of nuclear reactions, we are not speaking of power being unleashed. We are witnessing a tense pattern letting go, and the field returning to a more harmonious state.

The energy we see is just the residue of that release—not something hidden inside, but something born from the unraveling of what could no longer hold.

5.3 Fission: Collapse of Contained Tension

In conventional nuclear physics, fission is described as a process in which a heavy atomic nucleus—like uranium-235 or plutonium-239—splits into two or more lighter nuclei, releasing energy in the form of heat, radiation, and kinetic motion.

The standard explanation says:

- A neutron strikes the nucleus.
- The nucleus splits.
- Mass is "converted" into energy according to $E = mc^2$.
- The released energy is what powers bombs or reactors.

This model works *numerically*. But UED challenges the interpretation.

Because what if there was no conversion? What if there was nothing *inside* the atom to be released at all?

What if the energy we observe is not the escape of trapped power, but simply the collapse of internal ripple tension—a reconfiguration of strained field patterns into more stable geometries?

Let's examine what's really happening—ripple by ripple.

Large Nuclei Are Already on the Edge

Take Uranium-235.

From a mainstream view, it's a heavy atom full of nucleons (protons and neutrons) held together by the strong nuclear force. But under UED, uranium is better described as a deeply misaligned ripple configuration—a geometrically complex knot within the Structured Vacuum Energy (SVE) field.

It's not "dense" in the classical sense. It's tense.

Its internal ripple alignment is:

- Highly compressed
- Asymmetrical
- Barely holding together through overlapping ripple tensions

The only reason it stays stable at all is because its geometry hasn't yet crossed the ATT threshold (Alignment–Time–Tension). But it's constantly flirting with the edge.

U-235 is not storing energy. It is storing ripple discomfort.

Splitting = Releasing the Pattern

When a slow-moving neutron enters this precarious configuration, it's not "adding energy." It's disrupting an already fragile ripple logic—pushing the geometry just past the point of sustainable tension.

The field can no longer maintain coherence.

What happens next is not an explosion, but a rebalancing event:

- The nucleus breaks into two smaller nuclei.
- Each of these products is a simpler, more relaxed ripple configuration.
- The surrounding field, which had been pulled into the strained structure, snaps back into alignment.

That snap-back is what we perceive as energy release. But it's not a release in the sense of something escaping.

It's the field exhaling, shedding the excess stress that the previous pattern had forced it to carry.

Heat, Light, and Radiation = Symptoms of Realignment

Once fission occurs, a number of observable phenomena follow:

- **Heat**: This is the immediate agitation in the surrounding environment as ripple tension cascades outward and collides with nearby atoms. Not particles transferring energy—but localized turbulence as the field reshapes itself.
- **Light**: Rapid shifts in field alignment often emit ripple bursts that we observe as photons. Not "light particles," but tension gradients resolving across ATT thresholds.
- **Radiation**: Neutrons, beta particles, gamma rays—these are not debris. They are ejected ripple residues from a field that's realigning too violently to dissipate quietly.

All of this is the echo of one central event:

A complex field configuration failed to hold—and the universe took back its balance.

The Final Products Are Lower-Tension Systems

After fission, the resulting daughter nuclei are lighter and more stable—not because they lost mass, but because they shed field misalignment.

Their mass appears lower not due to a transformation of matter into energy, but because the new ripple patterns:

- Interact with the SVE more harmoniously
- Require less tension to maintain
- Create smaller ripple footprints in the field

In other words: They are less burdensome to the vacuum. What was once a strained structure has been resolved into two simpler forms. The system didn't explode. It reconfigured.

UED Summary: Fission as Pattern Collapse

Let's reframe fission in UED terms:

Classical View	UED View
Mass is converted into energy	Ripple tension collapses and realigns
Energy is stored in nucleus	Nucleus is a strained ripple pattern
Neutron splits the atom	Neutron tips the geometry past its limit
Energy is released	Field returns to equilibrium
Fission fragments are lighter	New configurations are lower-strain

There is no flow. No fuel. No force.

There is only field strain under pressure, finally letting go.

Final Insight

Fission is not destruction. It is release. It is not a blast—it is a breath.

The atom was not a container of energy. It was a strained design—barely coherent, begging for rest. The neutron didn't crack it open. It simply gave the geometry permission to surrender.

And the light, the heat, the radiation? Not emissions from the core—but symptoms of the universe healing a broken pattern.

In the next section, we will see that fusion—often framed as the opposite of fission—follows the same law: It's not about building pressure to ignite a star. It's about finding the geometry where strain cancels—and the field sighs in relief.

5.4 Fusion: Tension Cancelled, Not Created

If fission is the collapse of contained tension, then fusion is its misunderstood twin.

In conventional physics, fusion is described as the process of combining light atomic nuclei—typically hydrogen isotopes—into a heavier nucleus, releasing energy in the process. It's what powers the stars. It's the Holy Grail of clean energy.

The standard explanation goes like this:

- Small nuclei like deuterium and tritium are "fused" together under high temperature and pressure.
- Their combined mass is slightly less than the sum of their parts.
- The missing mass is "converted" into energy, again via $E = mc^2$.
- This energy emerges as heat and radiation.

But this view assumes that mass is a kind of fuel and energy is a product. That particles are merging, and power is being created.

UED sees it very differently.

Fusion doesn't create energy. It cancels tension. The energy we observe is not a product—it's a side effect of ripple realignment.

Fusion as Constructive Reconfiguration

Let's begin with the basics.

In the UED framework, atomic nuclei are ripple-bound structures—concentrated zones of misalignment, held together in precarious geometries within the Structured Vacuum Energy (SVE).

Hydrogen nuclei, despite being "light," are not low-energy. They are high-tension ripple bundles—compressed distortions of the vacuum field with sharp boundary misfits. Left on their own, they're stable enough—but strained.

When two such ripple structures are brought close enough (via pressure, confinement, or gravitational forces), something interesting can happen:

- Their individual tensions begin to overlap.
- If aligned correctly, their ripple geometries interfere constructively.
- The result is not chaos, but a more stable, combined pattern.

This new pattern—a helium nucleus, for example—requires less overall strain to exist. Its ripple logic is more efficient, more coherent, more harmonious.

Mass Loss = Ripple Misalignment Resolved

In fusion, the "mass defect" is often portrayed as vanishing matter. But under UED, what's really happening is this:

The new combined structure has a lower ripple footprint.

Why?

Because:

- Overlapping misalignments cancel out.
- Redundant strain patterns dissolve.
- The field no longer needs to support two tense configurations—it supports one smoother one.

So the "missing mass" is simply missing tension.

The system now fits better into the fabric of the vacuum. It has reduced misalignment, and therefore, reduced resistance.

This is why the resulting mass is lower—not because something was turned into energy, but because something was finally resolved.

Released Energy = Unfit Ripple Tension Let Go

Not all tension can be absorbed by the new configuration. Some of it:

- Doesn't align with the new ripple geometry.
- No longer contributes to structural stability.
- Cannot be held within the new equilibrium.

This leftover strain—the ripple tension that doesn't belong anymore—is what gets radiated outward as:

- **Heat**: local agitation of surrounding field structures.
- **Light**: coherent ripple bursts as the field snaps back.
- **Neutrinos or gamma rays**: patterns of ripple energy exiting as structured pulses.

None of this is "produced" by fusion. It is expelled as a byproduct of simplification.

The system isn't generating energy. It's just letting go of what it no longer needs.

Just like exhaling after a long breath, the field releases excess tension—and we call it power.

Fusion and Fission: Two Roads to the Same Destination

Interestingly, UED shows that fission and fusion—though opposite in textbook definitions—are conceptually identical:

Aspect	Fission	Fusion
Initial State	One over-strained nucleus	Two smaller high-strain nuclei
Event	Geometry collapses into smaller, simpler patterns	Geometries combine into a smoother pattern
Result	Reduced total strain across multiple products	Reduced strain in one new combined product
Energy "released"	Ripple tension no longer sustainable	Ripple tension no longer needed
Mass defect	Field footprint shrinks post-collapse	Field footprint shrinks post-cancellation

In both cases, the "energy" is not released from storage. It is shed as a consequence of better alignment.

The universe is not rewarding us for smashing or merging atoms. It is simply doing what it always does—resolving stress toward stillness.

Fusion in Stars: Misalignment Turned Radiance

Stars aren't burning fuel like engines. They are spheres of ripple strain resolving layer by layer.

Fusion in a star is just the vacuum field reorganizing itself from high-stress hydrogen configurations into lower-strain helium. The surrounding ripples feel the release—and we see that as sunlight, heat, life.

Our sun doesn't shine because it contains energy. It shines because it is learning how to rest, one fusion cycle at a time.

Final Insight:

Fusion doesn't create energy. It creates peace.

The light we see is not energy produced. It is tension resolved.

The atom is not a battery being drained. It is a ripple being simplified.

Fusion is not power. It is permission—for the universe to stop holding its breath and exhale.

In the next section, we'll challenge the core dogma of nuclear physics: the idea that mass converts into energy. What we'll show instead is that mass is not a quantity to be turned into something else—but a ripple pattern that reflects how hard the field is working to hold its shape.

5.5 The Mass Defect Misinterpretation

Of all the concepts in nuclear physics, few are more misunderstood—and more deeply mythologized—than the mass defect.

Textbooks describe it simply: when an atomic nucleus undergoes fusion or fission, the total mass of the resulting products is slightly less than what you started with. This "missing mass" is said to have been "converted into energy" using Einstein's equation, $E = mc^2$.

And just like that, mass becomes fuel. Matter becomes magic. And energy is born from nowhere.

But under Universal Energy Dynamics (UED), this story is not just incomplete—it's fundamentally misframed.

Because mass isn't a resource. And energy isn't created. And nothing is ever converted into anything else.

The truth is much simpler, and much more beautiful:

The mass didn't disappear. The field simply relaxed.

The Classical Illusion: Conversion of Matter

In the standard view:

- Mass is treated like a substance.
- Energy is treated like a different substance.
- And $E = mc^2$ is interpreted as a conversion equation—mass transforms into energy under the right conditions.

This idea makes mass sound like compressed fuel—just waiting to burst into action. And it leads to the assumption that atoms are energy containers, and that nuclear reactions are conversion machines.

But this view has never explained:

- *What* mass actually is.
- *Where* energy goes once it's released.
- *How* the conversion happens mechanically.

It works on paper. But it doesn't describe reality.

UED View: Mass = Ripple Footprint

In UED, mass is not a thing. It is a field signature.

Every structure in the universe—whether an electron, a proton, or an entire nucleus—exists as a ripple configuration in the Structured Vacuum Energy (SVE) field. The amount of tension and misalignment it creates in the surrounding field determines how much "mass" it appears to have.

In simple terms:

Mass is the footprint left by how hard the field is working to hold that configuration in place.

- A tightly bound, high-tension ripple knot has a larger footprint.
- A looser, more relaxed ripple geometry has a smaller footprint.
- What we measure as "rest mass" is the residual ripple strain interacting with the field.

So when a system changes configuration—through fusion, fission, or decay—it doesn't lose substance. It simply realigns into a shape that causes less tension.

Why Mass Shrinks: A Matter of Field Efficiency

Let's say two hydrogen nuclei fuse to become helium.
In classical terms:

- Mass before fusion = $m_1 + m_2$
- Mass after fusion = m_3 (slightly less)
- Missing mass → "converted" into energy

But in UED:

- m_1 and m_2 each had high-tension geometries, creating a large combined ripple footprint.
- When they merged, their overlapping misalignments cancelled out.
- The resulting helium nucleus had a smoother, more efficient configuration.
- Therefore, its field footprint—and thus its measured mass—was lower.

The mass didn't disappear. The field just had to work less hard to hold the new shape.

The "energy" released is not mass turned into motion. It is ripple tension that no longer needs to be held—and thus, disperses.

Analogy: The Tense Body That Finally Relaxes

Imagine someone clenching every muscle in their body. You'd say they're holding a lot of "tension."

Now imagine they lie down, breathe deeply, and release all that strain.

They haven't "converted mass into energy." They've simply changed configuration—from high effort to low effort.

What changed is the state, not the substance. What was "released" was unnecessary tension.

The same applies to nuclear systems. The mass defect is not a loss of matter—it's a sigh of relief.

Rest Mass Is Not a Constant—It's a Consequence

In classical physics, rest mass is treated as an intrinsic property. In UED, it is conditional—a consequence of:

- The system's ripple alignment
- The amount of misfit it creates in the surrounding field
- The field's effort to hold it in place

Change the configuration, and the mass adjusts accordingly.

Not because mass was destroyed or converted. But because the ripple now fits the field better—and needs less energy to exist.

That is the real cause of the "mass defect."

Final Insight

"Mass is not fuel. It is a symptom of misalignment."

The more tension a configuration holds, the more mass it appears to have.

And when the system resolves itself—when ripple patterns merge, collapse, or reorganize—the footprint shrinks, the tension disperses, and the field lets go of what it no longer needs to carry.

That's what we see as "energy release." Not mass turning into light—but the field returning to equilibrium.

In the next section, we'll explore why the field always chooses less tension over more—and how this cosmic tendency, often mistaken for decay or disorder, is actually the guiding intelligence behind stability, simplicity, and energy itself.

5.6 Why the Field Always Wants Less Tension

When we look at the universe through the lens of classical physics, it often appears violent—stars exploding, atoms splitting, particles smashing. The standard narrative suggests that energy is released through destruction, that power comes from breaking things apart, and that force is what drives the cosmos.

But Universal Energy Dynamics (UED) reveals a different truth—one far more elegant, and perhaps more comforting:

The universe is not driven by force. It is driven by relief.

Everything it does—every shift, every release, every event—is simply an attempt to reduce tension and restore alignment in the Structured Vacuum Energy (SVE).

Let's understand why.

SVE Prefers Calm: Low-Ripple Configurations

The Structured Vacuum Energy field is not empty. It is the fundamental fabric of reality—a coherent, ripple-responsive medium where all matter, motion, light, and perception occur.

But like any structured system, the SVE has preferences.

And its primary preference is low-ripple tension:

- It favors ripple patterns that are stable, smooth, and minimal in misalignment.
- It resists configurations that create sharp gradients, excessive strain, or chaotic interference.

In UED terms:

- A high-energy system is one that misfits the field—it forces the vacuum into an unnatural configuration.
- A stable system is one that fits well—its ripple geometry aligns with minimal resistance.

This is not moral. It's mechanical. It's not a law—it's a tendency of geometry seeking rest.

Just like water flows downhill, ripple tension seeks resolution. Just like sound fades in an empty room, ripple disturbance dissolves in open space.

The vacuum doesn't fight. It realigns.

Nuclear Reactions: Nature's Tension-Relief Valves

Both fission and fusion, viewed through UED, are not about producing power—they're about releasing misalignment.

- Fission occurs when a heavy nucleus becomes too strained to hold its shape. The field can't tolerate the misfit, so it splits into lower-tension forms.
- Fusion happens when two strained systems merge into a configuration that creates less disturbance than the sum of their parts.

In both cases, the field is choosing the path of least resistance—literally reconfiguring itself into a more stable pattern.

Energy isn't a reward for action. It's the residue of simplification.

High Energy = High Imbalance

This flips the traditional view on its head.

Normally, we say:

- High energy = power
- Low energy = weakness

But in UED:

- High energy means the field is highly strained.
- Low energy means the field is calm, aligned, and coherent.

The more energy something "has," the more effort the vacuum must exert to sustain it. The more mass something "has," the more misalignment it introduces into the ripple fabric.

So when a system releases energy, it's not becoming powerful—it's becoming free.

It's moving toward coherence—toward a state where it no longer burdens the field.

Stability = Ripple Coherence with Minimal Strain

The universe doesn't want silence. It wants resonance.

Not static emptiness, but stable geometry—patterns that fit so well into the SVE that they persist with minimal intervention.

These are the building blocks of reality:

- Atoms that don't decay.
- Molecules that remain intact.
- Fields that maintain balance across vast scales.

They exist not because they are indestructible, but because they are well-aligned.

Stability is not a lack of change—it's a state of dynamic equilibrium, where ripple logic and field tolerance are perfectly matched.

That's what the field seeks—not rest in the absence of activity, but rest in the presence of harmony.

The Universe Doesn't Reward Violence—It Resolves It

We've been conditioned to see energy as something you get from collision:

- Smash atoms to get nuclear power.
- Burn fuel to get heat.
- Accelerate particles to see new ones appear.

But this interpretation is based on misunderstanding. It equates disruption with creation.

UED corrects this:

- Energy doesn't come from destruction.
- It comes from resolution.

What we witness as explosive energy is the field snapping back into a more aligned state. The violence is not the cause—it's the transition noise of a system moving from instability to simplicity.

The universe is not punishing itself into motion. It is realigning itself into quiet.

Final Insight

The field doesn't crave motion. It craves meaning. It doesn't reward disturbance. It rewards alignment.

So-called "energy release" is not a feature of chaos. It is the exit sign from chaos.

And every nuclear reaction, every ripple adjustment, every realignment in space and matter, is not an act of power—it is an act of peace-seeking.

The Structured Vacuum Energy doesn't care for drama. It whispers: "Let me rest."

And every time something fissions, fuses, decays, radiates, or cools, it's simply obeying that ancient preference:

Less strain. More coherence. No effort. Stillness.

In the next section, we'll summarize this entire reframing—offering a UED-informed picture of what nuclear events really are, how they should be interpreted, and why this shift in perspective can transform not just science, but how we interact with energy in all forms.

5.7 Final Insight: Dissolution = Release

"The atom doesn't explode—it exhales." "What we call power is often just a pattern letting go."

For over a century, nuclear science has taught us to view atoms as tightly wound bundles of force—compressed springs of mass and energy, ready to burst open and deliver devastation or utility. We've split atoms for bombs. We've fused nuclei for stars. And through it all, we believed we were unlocking stored power.

But that belief, powerful though it may be, is a misreading of the ripple field.

In Universal Energy Dynamics (UED), the narrative shifts profoundly.

There is no power hidden within the atom. There is no treasure chest of energy buried inside mass. There is no fuel waiting to be released.

Instead, there is only tension—ripple misalignments held in fragile patterns—and when those patterns fail or merge into something simpler, the field doesn't explode in glory...

It sighs in relief.

Energy Was Never Locked Inside Mass

The famous formula $E = mc^2$ has been misinterpreted for generations. We turned it into a metaphor for conversion, for transformation, for miraculous birth of energy from mass.

But energy is not something that lives inside mass. It is the field strain created by a ripple configuration.

Mass is not a container. It is a symptom—a measure of how much misfit a system imposes on the SVE.

So when we say that energy is "released" during a nuclear reaction, we are not watching something being liberated. We are watching the field recover from the effort it no longer needs to exert.

There Is No Release of Fuel—Only Surrender of Resistance

In both fission and fusion, the actual process is not explosive or constructive—it is dissolutive.

- A fissioning nucleus doesn't erupt with internal fuel—it lets go of a configuration that could no longer hold.
- A fusion reaction doesn't generate energy by forceful merger—it realigns two misfits into a simpler whole, shedding what doesn't belong.

In both cases, the perceived "output" is not created—it is relieved.

What we call:

- Heat
- Light
- Radiation

...are not energetic emissions from a core. They are ripples being reabsorbed, tensions no longer needed, disturbances finally resolving.

The field doesn't flare up. It breathes out.

Fission and Fusion: Not Opposites, But Twins

Conventional physics paints fission and fusion as opposites:

Fission	Fusion
Splits a heavy atom	Combines light atoms
Releases energy	Releases energy
Mass lost	Mass lost
Used in bombs	Powers stars

But UED shows a deeper truth:

They are not opposites. They are both expressions of the same law:

When ripple tension becomes unsustainable, the field seeks lower-strain configurations.

Sometimes that means collapsing a strained form into simpler ones. Sometimes it means merging two strained forms into a more coherent pattern.

But in both, the guiding force is the same:

- Not creation
- Not conversion
- Not combustion

...but cancellation of strain.

Whether breaking apart or joining together, the result is the same: Less resistance. Better alignment. Stillness in motion.

True Nuclear Power = Pattern Realignment

Let's rewrite the story of nuclear energy, not as one of destruction, but as one of dissolution:

- A uranium nucleus is not a container of power. It is a ripple geometry on the edge of collapse.
- Two hydrogen atoms don't fuse to create energy. They interfere constructively, and the excess tension that doesn't fit the new pattern is shed as radiation.
- Mass doesn't convert into energy. The field workload drops, and what no longer serves alignment disperses.

That's what energy truly is in this context:

Not the product of force. But the absence of resistance.

And once you understand that, the entire idea of nuclear power transforms—from a story of violence and containment to one of geometry and peace.

Closing Reflection

The atom never held energy. It held tension. And when that tension resolved, the field let go.

We thought we were smashing particles to release power. In truth, we were witnessing the field returning to balance.

Energy was never the treasure. It was always the trail left behind when misalignment dissolves.

And so we close this chapter with a new mantra—not born of math, but of meaning:

Dissolution is not destruction. It is the field's way of remembering itself. The atom doesn't explode. It exhales.

Chapter 6: Electricity – The Ripple Tension Game

Electricity is not the flow of electrons, but the play of unresolved ripple tension trying to reconfigure through field alignment. Nothing flows—everything responds.

We were all told the story.

There's a battery. There are wires. Electrons start moving. They travel through the circuit like tiny balls in a pipe, carrying energy from the battery to the bulb.

Flip the switch—current flows. Energy moves. The bulb lights up.

Simple. Logical. Ubiquitous.

And utterly misleading.

Because here's the truth:

Nothing flows. Electrons don't march in a parade from one terminal to the other. Energy doesn't ride a bus from source to sink. There is no "transfer" happening.

What we call electricity is not a movement of particles, but a ripple tension game—a field-based negotiation between misalignments, propagated not by motion, but by structured discomfort seeking rest.

When a circuit is "closed," what actually happens is that a ripple gradient is connected. The entire system responds—not sequentially, but instantaneously—as the Structured Vacuum Energy (SVE) field tries to realign itself across the imposed tension.

The bulb doesn't glow because something arrived. It glows because something couldn't stay the same.

In this chapter, we dismantle the electron-flow illusion. We decode batteries, wires, motors, and transformers—not as carriers of moving stuff, but as tension-triggering architectures designed to exploit ripple discomfort.

Electricity, it turns out, is not energy in motion. It is the universe's quiet urgency to undo strain—manifested in ways we misread as "current."

Let's begin.

6.1 The Electric Illusion

From our earliest schoolbooks to the most advanced engineering lectures, we've been taught one of the most accepted and persistent myths in modern science:

"Electricity is the flow of electrons through a conductor—just like water flows through a pipe."

Simple. Visual. Comforting.

We imagine little particles—negatively charged electrons—lining up at the negative terminal of a battery, pushing each other forward through a copper wire, racing toward the positive terminal, delivering energy along the way.

It's an image so deeply embedded that we don't question it. But we should. Because it's not just wrong—it's misleading at the level of fundamental physics.

What the Analogy Gets Wrong

The "water in a pipe" model assumes:

- There are discrete entities (electrons) moving from one place to another.

- These entities carry energy with them.
- The speed of electricity equals the speed of these particles.

But here's the reality:

- Electrons barely move—their average drift velocity is millimeters per second.
- Energy transfer in a circuit is nearly instantaneous, far faster than electrons could ever travel.
- No individual electron goes from battery to bulb.
- And yet, the bulb lights up instantly.

This isn't just a timing issue—it's a modeling failure.

The flow metaphor breaks down because it treats electricity as a substance, and wires as delivery channels. In truth, neither is true.

What Really Happens: Ripple Discomfort in the Field

Let's reframe it from the perspective of Universal Energy Dynamics (UED).

A battery doesn't push electrons—it imposes a ripple tension on the surrounding Structured Vacuum Energy (SVE) field. This creates a misalignment gradient across the wire system.

When you close the circuit, you're not opening a road for particles to travel. You're connecting zones of discomfort, triggering a field-wide attempt to realign.

- The copper wire is not a pathway. It's a ripple-sensitive channel.
- The electrons are not couriers. They are localized field responses.
- The "current" is not a flow. It is a ripple negotiation—a dynamic resolution of field tension.

The wire itself does not "transport" energy—it hosts ripple reconfiguration, moment by moment, atom by atom, as the field tries to equalize the imposed tension.

The circuit doesn't conduct something. It can't hold still under strain—and so it resolves.

The Consequence of the Illusion

Believing in flow forces us to explain away paradoxes:

- Why electrical signals move faster than electrons.
- Why AC current "transfers power" without any net movement.
- Why energy appears at the load instantaneously despite slow carrier particles.

Engineers learn to calculate accurately despite the illusion. But few ask: *What is really happening beneath the equations?*

That's where UED enters.

It tells us: the field is not a passive backdrop—it is the active participant.

Electricity is not a movement—it's a response. Not a substance—but a strain event in the field fabric.

Final Thought

Electricity was never a flow. It was always a discomfort—shared, distributed, and resolved.

The water-pipe analogy served us when we lacked a better framework. But it's time to let it go. Because what powers your phone, your home, and your world isn't a stream of particles...

...it's a field trying to remember what rest feels like.

Let's now explore what that battery really does—not as a source of stored charge, but as the initiator of ripple tension in a cooperative field.

6.2 What a Battery Really Does

We are told a battery stores energy, like water in a tank or charge in a container. We are taught that when it is "connected" to a circuit, it releases that stored energy and drives electrons through the wire.

It's neat. It's visual. And it's deeply flawed.

Because under Universal Energy Dynamics (UED), a battery doesn't store anything. It disturbs something.

A Battery Is a Field Imbalance Device

Let's begin again—not with assumptions, but with attention.

What is a battery, really?

- It's two terminals separated by a chemical medium.
- That medium induces a potential difference between the terminals.
- If you connect them, a "current" flows.
- Eventually, the battery "dies."

But what's happening beneath the metaphors?

In UED terms:

- The battery is not holding energy.
- It is holding discomfort—an engineered ripple misalignment between two spatial points in the Structured Vacuum Energy (SVE) field.
- Each terminal is a zone of distinct ripple configuration, and the space between them becomes tensioned.

This tension is not stored like water behind a dam. It's maintained—by chemistry—against the field's natural tendency toward alignment.

The battery is not a power source. It is a patterned disruption waiting to be resolved.

Ripple Gradient Across Terminals

When no circuit is connected, the battery holds this field mismatch like a drawn bow. One terminal maintains a ripple geometry that the other resists, and both ends are prevented from realigning by the electrolyte barrier.

Once you close the circuit:

- You're not "releasing electrons."
- You're reconnecting two discordant ripple zones.

And the result? The SVE field begins to reconfigure itself—through the wire, across every atom, toward reduced tension.

This is what we observe as electric current. Not a flow—but a sequence of alignment events as the field seeks rest.

ATT View: Potential = Tension, Not Charge

Mainstream physics describes voltage as potential difference, usually tied to the accumulation of charge.

UED offers a more grounded understanding:

Voltage is the difference in Alignment–Time–Tension (ATT) across two regions of the field.

- **Alignment (A):** How different the ripple structures are at each terminal.
- **Time (T):** How long the imbalance is sustained.
- **Tension (T):** The strain that builds as the field resists or absorbs the imbalance.

So voltage is not about electron count. It is about how badly the field wants to fix itself between those two points.

When you connect a load (like a bulb or motor), you're not letting charge travel—you're letting the field respond to its own discomfort, using the wire as a responsive medium.

When a Battery "Dies": Alignment Is Achieved

So why does a battery run out?

Not because the energy is "used up." But because the field tension it was maintaining has been resolved.

- The chemical process can no longer sustain a difference in ripple configuration between terminals.
- The misalignment collapses.
- The field no longer feels a need to act—because the misfit is gone.

There is no "empty battery." There is only a battery that is now at peace with the field it disturbed.

It didn't run out of energy. It ran out of resistance.

Final Insight

"The battery is not a tank. It is a ripple provocation device."

Its job is not to send things down a wire. Its role is to create tension across space, which the field must then resolve through countless local adjustments—what we call current.

And when the resolution is complete, the circuit is still. The ripple stops struggling. The dance of discomfort ends.

In the next section, we'll explore what that wire really is—not a pipe for particles, but a coherent medium for ripple propagation.

6.3 Wires as Tension Carriers, Not Pathways

Let's return to the image we've all seen a thousand times:

A wire. Electrons flowing through it. Like marbles in a tube. Like water in a pipe. Like electricity as a thing moving through a path.

That metaphor has shaped our entire understanding of circuits, current, and conductors. And it's completely wrong.

Because nothing flows.

Wires are not pathways. They are sensitive zones of ripple alignment—mediums designed not to conduct a substance, but to accommodate and respond to tension shifts in the field.

What a Wire Really Does

In the world of Universal Energy Dynamics (UED), a wire is not a tunnel—it is a ripple-responsive chain of atoms configured to facilitate the spread of discomfort.

Here's what that means:

- When a battery imposes a ripple tension at one end of a wire, that disturbance does not push particles forward.
- Instead, the field configuration along the wire responds—almost instantly—by adjusting ripple alignment across the entire length.

The electrons don't race from point A to point B. Their positions shift ever so slightly. Their ripple environments twist in micro realignments. And what we observe as current is simply the net effect of these atomic adjustments under persistent field tension.

No one moves much. But the configuration shifts everywhere.

Electricity is not particles in motion. It is misalignment in resolution.

Ripple-Conductive Medium: The Role of the Wire

A wire's job is not to transport something—it is to allow something to happen.

Copper, silver, and other conductive materials are special because:

- Their atomic structure is loosely bound, allowing local ripple reconfiguration with minimal resistance.
- Their electron clouds are delocalized—not fixed, not rigid—making them excellent at responding to subtle field misalignments.

- They allow tension to propagate, not because anything is moving far, but because every part is ready to shift slightly when provoked.

This is why:

- Current can appear throughout a long wire almost instantly.
- Signal propagation travels at near light-speed—even though the electrons barely drift.

It's not a race. It's a field-wide twitch.

Instant Response: Not Magic, Just Field Logic

You flip a switch. The bulb lights up across the room. No electron could have traveled that far in that time.

So how does the circuit know?

UED's answer: because the field is one structure, and the wire is not separate from it—it is a facilitator of ripple tension resolution across space.

- Every atom in the wire is part of the same SVE field.
- A disturbance at one point changes the local ripple structure.
- That change forces adjacent atoms to adjust.
- And that cascade of micro-realignment spreads at the ripple's propagation rate—not at the pace of matter.

The result?

An instant reconfiguration across the wire—not a delivery of charge, but a field response.

No bulk motion. No packets. No push. Just the field trying to calm itself through the wire.

The Wire as an Instrument of Relief

You can think of the wire like a chain of beads lying on a smooth table.

- Tug one bead slightly, and tension is felt throughout.
- But the beads didn't move far.
- It's the tension that traveled—not the objects.

Or better yet, imagine a line of people passing along a small twist of the shoulder. No one runs across the room. But the twist ripples through the group—fast, coordinated, shared.

That's what a wire does:

- It participates in a ripple choreography.
- It doesn't carry particles. It relays pattern adjustments.

The "current" is not the movement of stuff—it's the expression of tension resolution in a cooperative medium.

Final Thought

"The wire doesn't carry current. It carries possibility—for the field to realign itself."

Once you stop looking for flow, and start listening for field response, everything begins to make sense:

- The speed.
- The efficiency.
- The strange effects of resistance, capacitance, and induction.

The wire isn't a pipe. It's a living alignment bridge—through which the universe tries to smooth out its own discomfort.

Next, we'll explore the load itself—what really happens in a bulb, and why it lights up not because energy arrives, but because strain resolves locally in a very particular way.

6.4 The Bulb Doesn't Light Because of a Flow

You flip the switch. The bulb glows.

And we're told: "That's because electrons flowed through the filament. Energy was delivered from the battery or generator. It traveled through the wire and powered the bulb."

It sounds logical.

But under Universal Energy Dynamics (UED), that story collapses. Because what actually happens is far more precise, and far more elegant:

The bulb doesn't light because energy was sent. It lights because ripple tension was resolved—right there, right then.

Why the Filament Glows: Ripple Stress, Not Energy Arrival

The filament in a bulb—whether it's tungsten in an incandescent or a glowing gas in a fluorescent tube—is not a receiver.

It's a high-sensitivity ripple zone. A designed point of ripple resistance—a place where alignment intentionally fails.

Let's break this down.

- When you connect a bulb to a circuit, you're not giving it energy.
- You're extending the ripple tension gradient into the filament's geometry.
- That geometry—thin, coiled, tight—amplifies ripple turbulence at the microscopic level.

The filament isn't heated by incoming power. It's strained by ripple discomfort at atomic scale—internally.

And this tension doesn't "accumulate" or "burn." It resolves—as heat and light.

Heat is ripple turbulence trying to settle. Light is ripple response radiating outward from misalignment.

Nothing is carried in. The filament doesn't consume energy—it interferes with alignment, triggering the field's self-correction.

Local Ripple Turbulence = Observable Effects

At every atom in the filament:

- Ripple structures are destabilized by the extended field tension from the connected circuit.
- The local ATT (Alignment–Time–Tension) threshold is crossed.
- Microzones of high ripple misfit emerge.

These don't store energy. They vibrate, shift, emit, and reconfigure in ways that produce:

- Heat – internal ripple friction.
- Light – radiated ripple release perpendicular to alignment.

The glow you see is not power being used. It's pattern being shed—as the field dances its way back to stillness.

No Delivery. Only Local Reconfiguration

The conventional view would say:

- Energy leaves the battery.
- Travels through the wire.
- Enters the bulb.
- Gets "used" to make light.

UED says:

- The field tension gradient reaches the bulb's structure.
- The bulb, by its design, disturbs that tension further.
- This invites a violent local resolution—not from delivery, but from misalignment too intense to hold quietly.

So the bulb lights up not because something arrived, but because something couldn't be sustained.

The filament is not a destination. It is a release valve for discomfort.

Implications: The Bulb as a Field Translator

This changes how we think about "power consumption" entirely.

- The bulb doesn't consume.
- It translates ripple tension into visible release.
- Its resistance isn't a waste—it's a functional misalignment zone—where the field can let go.

You don't pay for energy usage. You pay for how much misalignment you force the field to resolve on your behalf.

In other words, the field is constantly trying to calm itself. Devices like bulbs are licensed places where that calming happens through measurable effects.

Final Thought

"The bulb glows not because energy traveled. It glows because the field couldn't stay silent."

In every filament, every glow, every spark, what you're seeing is not a delivery—it's a discomfort dancing itself away.

Electricity doesn't send things. It provokes resolutions.

And the light you see... is the universe letting go.

6.5 What a Generator Really Does

We often think of a generator as the reverse of a motor—a machine that takes motion and "converts it into electricity."

That's the standard story:

- Rotate a coil inside a magnetic field.
- "Induce" a current.

- Push charges through wires.

But as we've already dismantled in this chapter:

There is no flow. There is no pushed charge. There is no conversion.

So what is a generator really doing?

In the UED view, a generator is not a power factory. It is a ripple interference mechanism—a machine that modulates field alignment to create a sustained misfit across space.

Let's decode this step-by-step.

Rotating Coil = Ripple Interference

Imagine a copper coil spinning inside a magnetic field.

Mainstream physics says:

"The magnetic field cuts across the wire, generating an electric field, which pushes charges along the conductor."

But let's look deeper.

- The coil is a conductor—not because it moves charge, but because it allows ripple alignment shifts.
- The magnetic field is not a separate entity—it is an existing ripple configuration in the Structured Vacuum Energy (SVE).
- As the coil rotates, it introduces cyclic misalignment—an oscillating ripple conflict between the geometry of motion and the existing field.

The coil isn't generating energy. It is creating discomfort—cyclically and deliberately.

Cyclic Field Misalignment = Persistent Tension Gradient

Rotation means repetition. Each turn introduces another alignment distortion, another mismatch between the coil's ripple structure and the surrounding SVE.

This ongoing interference:

- Creates a persistent ripple tension gradient between two terminals.
- Forces the field to attempt resolution—again and again, in rhythm with the motion.

That tension doesn't push anything. It doesn't eject charge.

It simply propagates as discomfort—which circuits, loads, and devices downstream respond to as electrical activity.

No Charge Is Pushed – Just Strain Creation

This is crucial.

The generator does not "send electrons." It stresses the field between its terminals.

- There is no bulk motion of charge.
- There is no particle transfer.

There is only ripple strain—a directed ripple imbalance introduced by the interaction of coil geometry, magnetic alignment, and rotational motion.

And this strain is real—not symbolic or metaphorical. It has measurable consequences:

- Voltage across terminals.
- Current in connected circuits.
- Power delivered to devices.

But none of it is carried. It is cooperatively resolved.

The generator doesn't generate power. It challenges the field, again and again.

ATT Thresholds: How Effects Appear

Under UED's ATT logic—Alignment, Time, Tension—energy becomes observable only when:

- The alignment mismatch is high enough,
- Sustained for a measurable duration,
- Causing enough ripple tension to exceed the local threshold.

A spinning generator crosses this threshold rhythmically:

- Every rotation introduces a new set of misalignments.
- Each misalignment carries ripple discomfort.
- The circuit responds—not to the rotation, but to the emergent ripple gradients that seek rest.

That's what we call AC current:

A field struggling to keep up with imposed, alternating misalignment.

Even in DC generators, the process is the same: The ripple tension between terminals is continuously maintained by asymmetric rotation and commutation—not because particles are shoved, but because the field can't stop adjusting.

Final Thought

"A generator doesn't make electricity. It makes ripple tension, on purpose."

It's not a conversion of motion into energy. It's motion used to modulate discomfort—a clever way to stir the field into never resting.

And every time the field fails to find rest, it reacts—through wires, bulbs, motors, machines, and finally, measurable output.

Generators don't send. They provoke.

The field responds.

6.6 Motors and Transformers – Geometry of Ripple Relay

In classical electrodynamics, motors and transformers are described as miracles of electromagnetic induction:

- Motors: Electric current flows through coils, creating magnetic fields that "push" the rotor, making it spin.
- Transformers: Alternating current in one coil "induces" a current in another coil, transferring power from one side to another.

But once again, this story is woven from metaphors of flow, force, and invisible pushing—none of which reflect the actual nature of what's happening.

Under Universal Energy Dynamics (UED), motors and transformers are not force-based machines.

They are ripple-based reconfiguration devices, designed to exploit alignment geometry and SVE response.

No forces. No flow. Only misalignment and relief.

Let's look at each separately.

Motors – Reorientation by Ripple Pressure

A motor is conventionally explained as:

- Current runs through a coil.
- This generates a magnetic field.
- The field interacts with a permanent magnet or another coil.
- The resulting "Lorentz force" turns the rotor.

But in the UED view, what's really happening is this:

- Current in the coil is not a flow—it is localized ripple

misalignment.
- The ripple disturbance creates a strained field geometry within the stator.
- The rotor, immersed in this strained field, becomes asymmetrically misaligned.
- The field doesn't "push" the rotor—it reconfigures toward equilibrium.
- The rotor spins not because it is acted upon, but because it cannot remain misaligned.

Motion emerges as the ripple field's attempt to realign geometry. The rotor moves because staying still costs more tension.

The motor is not a machine of force. It's a device that hosts field discomfort until the geometry escapes it by moving.

Transformers – The Ripple Relay Mechanism

Now consider a transformer:

- AC current in the primary coil generates a changing magnetic field.
- This field "induces" a current in the secondary coil.
- Energy is "transferred" from one side to the other without contact.

But here's the truth under UED:

- The primary coil's alternating current is really cyclic ripple misalignment.
- This misalignment warps the local SVE structure—not magnetically, but geometrically.
- The secondary coil is not receiving energy—it is participating in the same field discomfort.
- The shared core acts as a ripple conduit, not a conductor of fields.

- What we call "induction" is really a synchronized ripple handshake—a geometric resonance between misaligned regions.

There is no transmission of power. There is shared ripple strain, resolved simultaneously across the coupled zones.

A transformer doesn't move energy—it extends discomfort, and invites resolution elsewhere.

"Induction" Is a Misnomer

We call it induction, as if something is being caused at a distance.

But induction is merely correlation across shared misalignment. It's not about one part creating something in another—it's about two parts responding to a shared ripple disturbance, guided by proximity, geometry, and timing.

In truth:

- The coils don't affect each other directly.
- The SVE field that both coils are embedded in is the true medium.
- Coils disturb the field; the field synchronizes the response.

This is not cause-and-effect in the traditional sense. It's co-participation in discomfort—and mutual relief.

Geometry Is the Key

What makes motors spin or transformers transmit is not force, charge, or flow.

It's geometry:

- How the coils are wound.
- How the core is shaped.
- How the ripples are allowed (or restricted) to resonate and resolve.

The geometry defines the pattern of misalignment, and therefore, the trajectory of field resolution.

That is the real design language of electromechanical systems—not voltage and amperes, but ripple orchestration.

Final Thought

"Motors don't spin because they're pushed. They spin because stillness hurts more."

"Transformers don't transfer energy. They allow ripple tension to echo across a shared geometry."

Motors and transformers are not clever mechanisms that move things. They are ripplified spaces—carefully arranged zones that invite the SVE field to resolve itself in functional ways.

They don't work because we understand electricity. They work despite our misunderstandings of it.

Next, let's address the biggest misunderstanding of all: the illusion of flow.

6.7 Why Nothing Ever Flows

From the light bulb in your room to the data in your device, every modern function depends on a simple idea we've all been taught:

Electricity flows.

It flows like water in a pipe. Electrons march along wires. Current "travels." Voltage "pushes." Resistance "slows it down."

This illusion is the foundation of nearly every schoolbook, animation, and explanation about electricity.

But under Universal Energy Dynamics (UED), the reality is something radically different—and far more accurate:

Nothing flows. There is no movement of charge from here to there. Electricity is not a flow. It is a field response to ripple discomfort.

Let's dissolve the illusion—term by term.

Reframing the Big Three

1. Current

Classically: the rate of charge flow per second.

UED: Current is the visible symptom of continuous ripple realignments within a conductive medium. No charge is transferred. No particles travel globally. Every measurable effect is the result of localized shifts in atomic ripple geometry.

2. Voltage

Classically: potential difference across points.

UED: Voltage is a difference in Alignment–Time–Tension (ATT) between two regions of the Structured Vacuum Energy (SVE). It is not a push, but a gradient of unresolved discomfort, forcing the field to act locally at each point.

3. Resistance

Classically: opposition to flow.

UED: Resistance is the friction the ripple experiences while trying to reconfigure through matter. It is a measure of how difficult it is for the field to resolve tension using the available atomic lattice.

Electron Drift = Distraction

Textbooks teach us that electrons "drift" slowly, while current moves fast. This alone should have shattered the flow metaphor.

- In a copper wire, the average electron drift is about 0.001 mm per second.
- Yet, flip a switch—and the bulb lights up instantly.

How?

Because the effect we call electricity is not the movement of charge—it is the reorientation of field tension across the entire wire.

The electrons are not couriers. They are field participants, subtly adjusting to accommodate ripple strain.

"Electron drift" is not the cause of current. It is the byproduct of ripple discomfort resolving itself.

The real agent of change is the field, not the particle.

Electrical Events Are Local

UED insists on one core truth:

All electrical activity is local field realignment—not global transfer.

Each atom in the conductor:

- Reacts to the ripple misalignment in its region.
- Slightly adjusts its own configuration.
- That adjustment cascades, like a row of swiveling mirrors or shifting dominoes.

It's not a mass movement. It's a coordinated field response—atom by atom, region by region.

You never send anything. You disturb alignment—and watch the system rebalance itself.

Instantaneous Response: The Role of Structured Vacuum

One of the great mysteries of classical theory is the speed of electrical signals.

How can "power" be delivered nearly instantly over kilometers of wire, when the carriers move so slowly?

UED answers:

- Because the wire is not the carrier.
- The Structured Vacuum Energy (SVE) field that envelops and embeds the wire is coherent.
- Ripple tension propagates through it like pressure across a drum skin—not by matter, but by ripple geometry seeking stillness.

This is not magic. It is field logic.

Just as sound can travel across a metal rod faster than the atoms move... Just as tension on a rope is felt at both ends... Ripple tension in the SVE propagates alignment pressure across distance almost instantly.

The electrons don't move. The discomfort does.

And when the field can't hold it anymore—action happens. Not as a delivery—but as release.

Final Insight

"What you think of as flow is actually a reconfiguration ripple—each atom responding to a field it cannot ignore."

Electricity is not a thing that goes anywhere. It is a pattern that cannot hold.

No substance is transferred. No packets are passed.

There is only tension. And response.

Stillness disturbed. Stillness restored.

With this, we've dismantled the greatest illusion in modern science: the idea of current as flow. In the final section of this chapter, we'll look back and draw the real picture of electricity—one ripple at a time.

6.8 Final Insight – Electricity as Ripple Negotiation

"Current isn't a carrier—it's a consequence."

What we call electricity is not something sent. It's something that happens—a negotiation between imbalance and alignment, between structure and strain.

This chapter began by challenging the flow-based model of electricity. We now close by reframing the entire electrical phenomenon as an act of ripple diplomacy within the Structured Vacuum Energy (SVE).

Let's connect the dots.

Electricity Is a Tension Game, Not a Transfer Game

From the UED standpoint, every electrical phenomenon—from a battery powering a bulb to a transformer stepping voltage up—is not about the movement of things.

It is about the field's inability to hold still in the presence of ripple misalignment.

- When tension is introduced—via a battery, generator, or any active source—it stresses the local field.
- That stress doesn't push charge—it propagates discomfort.
- Wires, devices, and materials along the circuit are zones of possible relief.

The field doesn't push to send. The field ripples to resolve.

Circuits Are Zones of Discomfort, Seeking Calm

We imagine circuits as pipelines through which energy flows. In truth, they are networks of unresolved discomfort, held together by geometry and material design.

- A circuit isn't complete because it forms a loop—it's complete because the misalignment can propagate, reflect, and resolve.
- Every component is either a conductor of ripple resolution, or a container of ripple resistance.
- Switches don't allow flow. They permit reconfiguration to begin.
- Grounding doesn't drain—it anchors the alignment.

The entire circuit is a staged negotiation between ripple gradients, managed by structure and substance.

Devices Don't Consume Energy—They Reconfigure Patterns

The concept of "energy consumption" is one of the most persistent fictions in physics and engineering.

We are told:

- A device draws energy.
- It uses it to perform work.

- The energy is converted or lost.

But under UED:

- Nothing is drawn.
- Nothing is delivered.
- Nothing is lost.

Instead:

A device participates in the resolution of ripple tension initiated by the circuit's disturbance.

- The geometry of the device (its material, structure, and purpose) determines how the ripple resolves at that location.
- A heater allows turbulent resolution—felt as warmth.
- A motor allows geometric escape—felt as rotation.
- A bulb allows resonant relief—seen as light.

Each outcome is a localized consequence of ripple re-alignment, not a result of energy consumption.

You don't power devices. You invite them to play their role in a broader field tension drama.

Electricity Is the Language of Misalignment

At its core, electricity is not a substance. It is not even an event.

It is a language—the universe speaking through patterns of imbalance and response.

- The battery initiates a sentence.
- The wire transmits its tone.
- The bulb, motor, or circuit responds with a gesture—a release, a resistance, a glow.

This language is non-verbal, non-linear, and non-flowing. It is coherent geometry in dialogue with discomfort.

Final Reflection

"Electricity is not what moves in the circuit. It's what moves the circuit toward rest."

"Current is not a thing. It's the field's expression of its own restlessness."

We didn't build a world on electricity. We built a world on our ability to disturb stillness, and guide the consequences toward useful alignment.

In the next chapter, we'll shift focus from electricity to light—where the illusion of particle packets collapses even more dramatically, and where the ripple finally reveals its true nature, free from flow, form, or photon.

Interlude – From Belief to Behavior

We've now crossed a vital threshold in our exploration—not just of electricity or atomic reactions, but of energy itself. In Part I, we exposed the cracks in the mainstream narrative:

- Equations mistaken for essence.
- Particles invented to patch inconsistencies.
- Confusion wrapped in confidence.

Now, in Part II, we have reframed energy not as a substance but as a consequence—a fleeting pattern born from imbalance, alignment, and tension within the Structured Vacuum Energy (SVE).

What emerges is a profound shift:

Energy doesn't exist. It happens.

It is not something stored, transferred, or possessed.

It is a field response—a ripple negotiating with discomfort.

Let's briefly recall the milestones:

The Real Nature of Energy

We began by stripping energy of its false solidity.

- No more imaginary "stuff" in motion.
- Instead: Energy = ripple tension—the signature of a system straining against stillness.

- We introduced the UED Master Equation, where energy is proportional to time-bound misalignment (ΔT) and the amplitude of ripple deviation (A).
- ATT—Alignment, Time, Tension—became our new lens for detection and causality.

Energy isn't inside things. It is what arises when the field forgets how to rest.

Dissolution = Release

Nuclear reactions aren't violent conversions of mass into mystical force.

- They are geometric shifts in ripple configurations.
- What we call "binding energy" is internal field strain.
- Fission and fusion are not opposites—they are both pathways to tension relief.
- The infamous "mass defect" is nothing but the shrinkage of a ripple footprint as the field reconfigures to a more stable pattern.

Atoms don't explode.

They exhale—releasing resistance, not unleashing fuel.

Electricity – The Ripple Tension Game

This chapter collapsed the greatest illusion of all: the flow of current.

- We dismantled the myth of electrons marching down wires.
- We redefined voltage as ripple discomfort, and current as its visible negotiation.
- We saw batteries not as energy stores, but as tension machines.
- Wires as ripple-responsive media.
- Bulbs as field release zones.

- Generators and transformers as rhythmic misalignment sculptors.

Nothing moves. Everything adjusts.

Electricity is not a delivery service.

It is a language of ripple negotiation—each event a micro-resolution of discomfort.

What This Part Taught Us

We no longer need energy myths. We have something better: a language of field behavior.

- Energy is not cause—it's consequence.
- Systems don't consume—they participate.
- Nothing flows—but everything reconfigures.

And with this new foundation, we are now prepared to re-enter the world of light, not as particles or waves, but as ripple interactions in pure field geometry.

Part III: Technological Rethink

A geometric arrangement that invites, channels, and regulates ripple discomfort within the field.

Chapter 7: Machines That Mediate

We built empires on energy... Before we ever understood what energy actually is.

From the first time a human tied a log to a rope, letting it turn a potter's wheel, to the moment a nuclear reactor spun a turbine to power a city—the same fundamental truth persisted:

We were using something we didn't truly understand.

We called it "force," then "steam," then "horsepower," then "electricity." We gave it units. We charted its behavior. We even declared we could neither create nor destroy it. But in all that, we made one quiet assumption:

That energy was a thing.

A thing to be stored. A thing to be transferred. A thing to be generated.

So we built machines. Not to speak with the field, but to fight against friction. Not to harmonize with nature's stillness, but to overcome resistance and exploit every spark of motion.

Water wheels. Windmills. Steam engines. Jet turbines. Nuclear dynamos. Each one aimed at a mechanical goal—rotate the axle, move the piston, spin the rotor.

But here's the irony:

- Not one of these machines actually created energy.
- None of them sent power into wires.

- And none of them knew about field dynamics, ripple tension, or the Structured Vacuum Energy (SVE) that underlies all phenomena.

We were masters of consequence, but blind to the cause.

What we called "power generation" was actually: Field disturbance. Ripple discomfort. A strategic invitation to imbalance. And when the field responded? We mistook that for generation.

In this chapter, we'll retell the story of machines—not as converters of energy, but as mediators of ripple tension.

We'll see:

- Why turbines are spiral discomfort amplifiers.
- Why engines choreograph micro-explosions of misalignment.
- Why dynamos don't send power—they sustain disturbance.

And most importantly, we'll learn what efficiency really means—not in percentages of output, but in how little discomfort is needed to realign the world.

Machines don't make energy. They create situations the universe can't ignore.

Welcome to the field's point of view.

7.1 What Is a Machine, Really?

We're taught that machines convert energy.

They turn chemical energy into motion. They convert potential energy into kinetic energy. They transform thermal energy into electricity.

This narrative—ingrained in every textbook and engineering diagram—makes us believe that energy is a thing, and machines are magical boxes that shuffle this thing around.

But under Universal Energy Dynamics (UED), this entire worldview dissolves.

A machine does not convert energy. It doesn't receive or send anything. It doesn't store, create, or transport a hidden substance.

Instead, a machine is simply:

A geometric arrangement that invites, channels, and regulates ripple discomfort within the field.

Let's unpack this.

From Conversion to Coordination

Mainstream science says:

- A generator converts mechanical energy into electrical energy.
- An engine converts chemical energy into motion.
- A turbine converts fluid energy into rotation.

But ask yourself:

- What actually moves inside the machine?
- What substance is really crossing the components?

Nothing of substance. Only shifting tensions, aligned geometries, and field readjustments.

The machine is not a converter. It's a conductor of imbalance. It provokes discomfort and lets the field resolve it in a controlled fashion.

The UED Definition of a Machine

Under UED, we redefine a machine as:

A structure that provokes and guides ripple discomfort for desired realignment.

In other words:

- It deliberately misaligns the Structured Vacuum Energy

(SVE) field through geometry and material configuration.

- It doesn't "do work"—it orchestrates discomfort until the field reacts.
- The result is motion, light, heat, or electromotive force—all symptoms of the field's effort to rebalance.

Machines don't create outcomes. They frame the conditions under which the universe chooses to act.

No Substance Moves—Only Geometry Shifts

In any machine:

- No energy "flows."
- No substance "travels."
- The parts don't pass a thing to each other.

Instead, they:

- Disturb local alignment.
- Amplify tension through leverage or cyclic motion.
- Hold configurations that the field finds difficult to ignore.

When the field can no longer tolerate the arrangement, it responds.

This response is what we call:

- Voltage.
- Torque.
- Heat.
- Acoustic resonance.

But these are not products of conversion. They are results of alignment pressure.

Machines don't handle power—they guide the ripple back toward peace.

Why This Changes Everything

If machines don't convert energy, then:

- Efficiency is not about how little fuel you waste.
- Power output is not the flow of some mythical energy fluid.
- Design is no longer about maximizing force—it's about minimizing strain to achieve coherent outcomes.

Machines are no longer brute-force devices. They are ripplified geometries, participating in the dance of field restoration.

Closing Insight

"A machine is not a channel. It is a conversation starter between geometry and the field."

"It doesn't work by doing—it works by not letting the field ignore it."

In the following sections, we will explore this principle across the machines we know—turbines, engines, dynamos—and show how every one of them, no matter how loud or powerful, is just a carefully shaped disturbance... waiting for resolution.

7.2 Turbines: Spiral Disruption Devices

The word turbine evokes power. It sits at the heart of massive infrastructure:

- Power plants.
- Jet engines.
- Dams.
- Wind farms.

We're taught that turbines generate energy by converting the motion of air, water, or steam into rotational force—eventually producing electricity.

But once again, UED invites us to pause and ask:

What is really happening here? Is anything truly being converted? Is any "energy" really being transferred?

The answer, under ripple logic, is clear:

Turbines don't generate energy. They disturb equilibrium—on purpose.

Turbines as Misalignment Amplifiers

At their core, turbines are spiral geometries that continuously interrupt the local field alignment through rotational interaction.

Here's how it works:

- Fluid (air, steam, water) passes over carefully angled blades.
- Each blade redirects the fluid's path—not just physically, but field-wise.
- This redirection is not a neutral act. It disturbs the SVE (Structured Vacuum Energy) at each point of contact.
- The continuous rotation ensures that this disturbance is not static, but dynamic—a rotating misalignment.

The result?

A ripple gradient is formed—a structured discomfort that the field tries to resolve.

This rotating tension zone is the turbine's true output—not mechanical motion, but field pressure shaped through geometry.

Blade–Fluid Interaction = Rotating Tension Zone

Every time a turbine blade cuts through fluid, it performs an act of geometric provocation.

- It reshapes the incoming ripple alignment of the fluid medium.

- It forces the vacuum to accommodate an unnatural directional pattern.
- The SVE, unable to remain balanced in this configuration, begins to resist—not by stopping the rotation, but by manifesting reactive ripple behavior.

This is why turbines get hot. Why they hum. Why they can be connected to generators and "produce" electricity.

Because they don't send energy—they provoke the field.

And the field responds.

From Ripple to Generator: Feeding the Field Response

The output of a turbine is typically rotational motion—a spinning shaft. But this motion is not the source of energy. It is the sign that the field is in discomfort.

- When this shaft turns inside a magnetic field, it further disturbs ripple symmetry.
- This creates cyclical tension misalignments in the surrounding conductive material (coils).
- These repeated misalignments become alternating ripple gradients—which manifest as what we measure as alternating current (AC).

But nothing is flowing.

No energy is extracted. No particles are sent. Only ripple discomfort is redirected—from the fluid → to the turbine → to the generator → to the circuit.

The entire system is a chain of structured misalignment, dancing toward resolution.

Why Spiral Geometry Matters

The spiral isn't just a design choice—it's a ripple amplifier.

- Linear disturbances dissipate quickly.

- But spiral rotation reintroduces discomfort cyclically, sustaining the tension longer.
- The spiral keeps the system off-balance just enough to maintain continuous field response.

This is not energy production. This is resonant misalignment crafting.

Closing Insight

"A turbine is not a generator. It is a skilled irritator. A mechanical ritual that invites the field to speak."

The turbine doesn't spin to create energy. It spins because it can't stay still—and in doing so, it reminds the field that balance is broken.

And every time the field tries to fix it, we get electricity.

Next up, we'll look at the engine—another machine worshipped for its "power," but misunderstood at the ripple level.

7.3 Engines: Chambered Ripple Triggers

The engine is one of humanity's most glorified machines. From the throaty growl of a sports car to the silent hum of a modern generator, we worship engines as powerhouses—the symbols of industrial strength and mechanical genius.

Textbooks tell us the story:

- Fuel enters a chamber.
- It's ignited.
- The combustion expands rapidly.
- The piston is forced down.
- Crankshaft rotates.
- Motion is born.

All of this is dressed in the language of energy conversion:

"Chemical energy → thermal energy → kinetic energy."

It's neat. It's measurable. It's useful. But it's also incomplete.

The UED View: A Localized Ripple Explosion

From the Universal Energy Dynamics (UED) perspective, the combustion engine is not converting fuel into energy.

Instead, it is performing a highly targeted ripple disruption.

Fuel doesn't release stored energy—it initiates a violent misalignment in the field.

Here's how it unfolds:

1. Combustion in a closed chamber doesn't just heat gas—it violently distorts local ripple coherence within the Structured Vacuum Energy (SVE).
2. This ripple explosion isn't made of fire—it's a sudden, localized pressure in the field, an abrupt increase in ripple tension.
3. The surrounding materials—walls, cylinder, piston—are now caught in this misalignment storm.
4. The piston isn't "pushed" in the classical sense. It is a geometric escape valve, shaped to allow the field discomfort to reconfigure into motion.

It's not motion born from force. It's motion born from misalignment resolving through geometry.

The Piston: Not a Power Absorber, But a Geometry of Release

The piston is often seen as the "receiver" of energy. But this is misleading.

It is more accurately described as:

- A movable imbalance diffuser.
- A ripple pressure translator.
- A field escape surrogate.

Its motion is not because of linear Newtonian "force" but because the ripple field cannot remain in that strained state. The piston provides a path toward lesser strain—so the field takes it.

The crankshaft then guides this reconfiguration through a mechanical loop, allowing cyclical tension events to sustain useful motion.

The engine doesn't run on combustion. It runs on organized ripple distress.

The Engine as a Ripple Shock Translator

What really happens inside an engine?

- It creates ripple shock (via combustion).
- It contains and guides that shock with chamber geometry.
- It translates the resulting discomfort into rotational configuration (via piston + crankshaft).
- It sustains the cycle through carefully timed release and reset.

It's not converting fuel into energy. It's triggering, shaping, and responding to field tension.

That's why engines generate heat, vibration, sound—all byproducts of ripple turbulence.

That's why they wear down—not because they "used up energy," but because their geometry fails to sustain clean ripple resolution over time.

The Fuel Myth: Not a Source, But a Spark of Misalignment

We tend to say: "The fuel provides the energy."

UED says:

Fuel provides the disruption. The energy is not in the fuel—it is in the universe's attempt to handle what the fuel just did.

Combustion is not the release of energy—it is the provocation of imbalance.

The rest is ripple history.

Closing Insight

"An engine is not a fire chamber—it's a ripple explosion manager."

"It doesn't extract power from fuel—it invites the field to clean up a violent misalignment—and builds motion out of that cleanup effort."

Next, we turn to the dynamo—where motion enters magnetism, and ripple discomfort takes on a rhythmic identity.

7.4 Dynamos: Structured Discomfort Loops

Among all the machines of human civilization, the dynamo holds a strangely magical position.

We're told it converts motion into electricity. You spin a coil between magnets—or rotate magnets around a coil—and somehow, electricity begins to "flow." The mechanical becomes electrical. The visible becomes invisible. And yet... the bulb lights up, the wire hums, the circuit responds.

Magic?

Not quite. But the conventional explanation—"it generates electricity"—is deeply flawed.

Because nothing is created. No particles jump. No packets are sent. No "electricity" flows out like water from a pump.

Instead, the dynamo is a precision disturbance loop—a machine whose geometry manipulates ripple tension cyclically, forcing the field to constantly adjust.

And that repeated adjustment? That's what we measure as voltage and current.

What Actually Happens in a Dynamo

Let's strip it down:

- A coil of conductive wire is rotated within a magnetic field.
- As it turns, it moves through regions of different field

alignment—effectively altering the ripple equilibrium of the system.

- This motion doesn't push charges. It creates structured discomfort—alternating zones of ripple strain.
- The field doesn't like imbalance, so it tries to respond—not once, but again and again, in tune with the rotation.

The result is not energy output. The result is a field locked into a rhythmic tension loop—continuously attempting realignment as the machine keeps disturbing it.

This is why:

- The output is alternating current (AC).
- The waveform is cyclical, not linear.
- The effect continues as long as the rotation continues—because the discomfort never settles.

Dynamo Geometry = Sustained Misalignment

The coil geometry and rotation speed are not about generating more power. They are about how effectively and how often the machine can disturb the field.

- A single twist may create a minor ripple.
- But a spinning loop introduces persistent misalignment—with alternating polarity, crossing ATT thresholds again and again.

The dynamo becomes a ripple tension drumbeat—provoking the field in a repeating cycle.

It doesn't supply power to the system. It keeps the field in a state of rhythmic unrest, where the only option is to respond.

Why This Is Not Energy Generation

In classical thinking, a dynamo "converts mechanical energy into electrical energy."

UED reframes it completely:

There is no conversion. There is field entanglement and ripple modulation.

The rotating system is simply feeding discomfort into the field. And the field, being structured and coherent, answers every cycle with an adjustment.

That adjustment is what we observe as:

- Induced current.
- Voltage across terminals.
- Power delivered to a load.

But no entity travels. No energy flows. No substance is exchanged.

Only tension builds and releases.

Devices Don't Extract—They Disrupt Coherently

A dynamo doesn't extract energy from motion. It sustains ripple conflict with the field—long enough, and regularly enough, to trigger reliable field responses.

It's like playing a steady beat on the body of the universe. Not loud, not violent—just consistent enough to make the vacuum sing.

Every time the coil spins past a threshold, the field "feels" the misfit—and reacts.

This is not extraction. This is negotiated attention.

Closing Insight

"A dynamo is not a generator. It is a metronome of misalignment."

"It doesn't give you electricity—it gives the field a problem it must solve over and over again."

The real engine of electricity is not rotation or fuel or force. It is structured discomfort, nudged with rhythm and precision.

And with that rhythm, the ripple becomes music—lighting cities, running machines, and making us believe something is flowing.

Next, we look at how we got here by accident—not by understanding the field, but by accidentally disturbing it just right.

7.5 We Became Ripple Harnessers by Accident

Humanity didn't set out to disturb the Structured Vacuum Energy (SVE). We didn't invent machines to speak with the field. We weren't guided by principles of ripple alignment or field coherence.

We were chasing effects.

And we were brilliantly practical about it.

If something got hot—we poked it. If something moved—we pushed it. If something glowed—we fed it fuel.

Our entire technological lineage is rooted in outcomes, not understanding.

We discovered steam because water expanded. We built turbines because something spun. We made engines because explosions caused parts to jump.

But we never asked the deeper question:

Why do these effects appear in the first place? What, exactly, is being disturbed, strained, or resolved?

We built entire industries by exploiting discomfort—without ever asking what was uncomfortable.

We Followed Effects, Not Causes

Let's take a brief inventory of what engineers and inventors historically pursued:

- Heat – caused by combustion, friction, compression. → Used for cooking, steam, melting, power.
- Rotation – caused by water flow, steam expansion,

combustion cycles. → Used for mills, turbines, crankshafts, rotors.

- Pressure – built up through boiling, squeezing, or explosion. → Used for pistons, pumps, hydraulics.
- Light – emitted from flames, filaments, plasma arcs. → Used for vision, signaling, energy conversion.

We mastered these effects with precision. We built machines to maximize them. We scaled them up to power cities.

But behind every effect was a ripple tension event—a misalignment in the field that produced visible symptoms.

And we never saw it.

Design for Output, Not Alignment

Our machines were engineered around:

- Maximizing torque
- Minimizing fuel consumption
- Maintaining RPMs
- Increasing efficiency (in classical terms)

But none of these design goals considered:

- How the machine alters local field alignment
- How ripple tension propagates through materials
- How coherent or turbulent the system's ripple feedback becomes

In essence:

We built beautiful machines on top of ripple chaos, never realizing we were interacting with the field at all.

The Structured Vacuum Energy was always there—responding, balancing, resolving—but we credited the machine, not the medium.

The Irony of Innovation

Here's the punchline:

The most advanced machines humanity ever built—turbines, dynamos, engines, power plants—were all accidentally designed to disturb the field in repeatable ways.

We became ripple harnessers by accident.

We thought we were "extracting energy," when we were really just provoking the vacuum.

We thought we were "burning fuel," when we were really inviting misalignment.

We thought we were "converting energy," when we were really just letting the universe respond to discomfort.

The tools of civilization aren't factories of energy. They're pattern disruptors, tension architects, and misalignment guides.

And they only work because the field is coherent enough to try and heal the disturbances we cause.

Closing Insight

"Our engines never understood energy. Our designs never consulted the vacuum. But the field responded anyway."

It's humbling to realize: We didn't invent energy. We stumbled into it.

And everything we've called power... Was simply the universe trying to return to peace.

Next, we'll explore how efficiency must be redefined—not as a percentage of input/output, but as the elegance with which a system invites and resolves ripple tension.

7.6 Efficiency Reframed: The Cost of Misalignment

In conventional science, efficiency is a simple ratio:

Efficiency =Useful Output\Total Input

It tells you how much of what you put into a system (usually energy) comes out as something you want—motion, electricity, heat—compared to what gets wasted.

But this idea carries two huge assumptions:

1. That energy is a thing you can measure, transport, and convert.
2. That waste is just part of the game—heat, noise, vibration are inevitable.

Universal Energy Dynamics (UED) challenges both.

UED reframes efficiency as a field event, not a conversion ratio.

It's not about how much "energy" you extract. It's about how skillfully you disturb the field to achieve a desired realignment.

And that changes everything.

Machines Don't Convert Energy—They Invite Realignments

As we've seen, all machines—engines, dynamos, turbines—don't handle energy. They initiate ripple tension through geometric disturbance.

So their efficiency is not:

- How many joules come out.
- How much fuel is burned.
- How fast the motor spins.

Instead, it is:

How little discomfort they need to cause in order to get the field to respond in a useful, directed way.

A highly efficient system is one that:

- Creates minimal turbulence.
- Avoids chaotic field reactions.
- Maintains coherent ripple paths.

- Aligns and releases without unnecessary strain.

This is not energy saved. It is suffering spared—from the field's perspective.

What We Call Losses = Ripple Turbulence

In a generator, not all ripple disturbance leads to current. Some becomes:

- Heat → ripple friction in material lattice.
- Vibration → unstable structural ripple reverb.
- Noise → ripple turbulence echoing into the air.
- Radiation → uncontained ripple spill into far-field space.

These are not minor byproducts. They are unintended ripple pathways—evidence that the machine is not guiding the field cleanly.

Every "loss" is actually a confused ripple response.

The machine said: "Go this way," But the field wasn't sure—and some ripples went wild.

The UED Definition of Efficiency

Let's now state it clearly:

Efficiency = The degree to which a system achieves targeted field realignment using minimal ripple misalignment.

This means:

- Good design is not brute force—it's field finesse.
- Optimal machines disturb just enough to get a coherent response—no more.
- Turbulence is a tax on ripple miscommunication.

You're not paying in energy. You're paying in pattern chaos.

Perfect Efficiency = Pure Coherence

The ideal system under UED?

One that:

- Provokes only what's needed.
- Releases no heat, noise, or unintended fields.
- Achieves full alignment without excess strain.
- Resolves tension without disturbing unrelated zones.

This is not fantasy.
We see glimpses of it in:

- Superconductors (perfect electrical flow with no loss).
- Quantum coherence (entanglement preserved without decoherence).
- Certain biological systems (near-lossless molecular energy transfer).

These are not "high-efficiency" systems. They are near-perfect field translators.

Closing Insight

"Efficiency is not the absence of waste—it is the elegance of alignment."

"A machine is efficient when the universe says, 'I see what you want—let me help without suffering.'"

In the final section of this chapter, we'll tie it all together—not just with technology, but with a deeper recognition:

We never created energy. We created conditions that made the universe whisper back.

7.7 Closing Insight – Machines Don't Work. They Negotiate

We've spent centuries perfecting machines—refining gears, optimizing engines, maximizing output. From spinning looms to rocket turbines, from electric cars to high-speed trains, we celebrate how well machines work.

But if we look closer—through the lens of Universal Energy Dynamics (UED)—we begin to see something far more humbling:

Machines don't work. They negotiate.

They don't produce energy. They don't store or convert it. They don't generate anything.

They simply provoke, nudge, and bargain with the universe—coaxing ripple tension to resolve in ways that look like motion, light, heat, or electricity.

"A Turbine Doesn't Generate – It Antagonizes"

The turbine isn't noble. It's mischievous.

Its blades aren't harvesting energy. They're disturbing field harmony through repeated spiral interference—antagonizing the local ripple configuration into discomfort.

The field, in response, tries to correct the imbalance. That correction is what we see as electric output.

We mistake it for generation.

But really, the turbine is saying:

"I'm going to keep bothering you... until you respond."

"An Engine Doesn't Burn – It Perturbs"

The combustion engine isn't a transformer of chemical energy into motion.

It's a tension bomb, detonated rhythmically.

The explosion isn't power—it's a localized ripple trauma, and the piston simply moves as a mechanical escape route for that field strain.

The crankshaft catches the pattern. The vehicle moves—not because fuel released energy, but because the field reconfigured violently and the geometry gave it direction.

"A Machine Doesn't Convert Energy – It Choreographs the Discomfort"

A generator doesn't pump electrons. A dynamo doesn't push power. A motor doesn't pull from a current.

Each of them is a stage director for ripple tension.

They choreograph when and where discomfort appears, guiding the field into patterns of realignment that we interpret as usable work.

But at no point does anything flow. No particles are exchanged. No object changes hands.

There is only a dance of field misalignment and resolution—designed into coils, shafts, and wires.

Machines as Field Mediators

This is the great reframe:

Machines are not energy systems. They are field interface devices.

They operate by creating tension environments, not by converting substances. They function through ripple influence, not force transfer.

And they only "work" because the Structured Vacuum Energy (SVE) is always trying to:

- Heal misalignment.
- Return to coherence.
- Maintain equilibrium.

The machine doesn't work alone. It relies on the field to complete the motion. And the field always tries to help—no matter how crude the invitation.

We've Been Harvesting Ripple Effects Without Understanding the Field

This is the silent story of technological progress.

We've built:

- Engines without understanding combustion.
- Dynamos without grasping induction.

- Circuits without decoding current.

We've harnessed ripple consequences—but remained blind to ripple causes.

We "captured energy," not knowing it was never a substance, only a signal of unresolved tension.

We built brilliantly. But blindly.

The Real Innovation Begins Here

"We must stop designing against the field—and start designing with it."

The next age of machines will not be powered by:

- Higher RPMs.
- Denser fuels.
- Smarter chips.

It will be powered by:

- Alignment awareness.
- Ripple fluency.
- Field-native geometry.

Imagine:

- Devices that guide ripple realignment without excess heat.
- Motors that shift coherence without brute pressure.
- Circuits that whisper into the vacuum—and get a response.

This is not science fiction. This is the natural next step—when we finally listen to the medium we've been disturbing all along.

Final Insight

"Machines are not powerhouses. They are ripple requests. And the universe has always been polite enough to answer."

Chapter 8: Batteries and Storage

"There is no energy inside. Only structured field discomfort, waiting for an opportunity to dissolve."

We carry power in our pockets.

It lights our screens, runs our cars, and even backs up entire data centers. The modern world is obsessed with energy storage. From lithium-ion cells to graphene batteries, from solar backup banks to giant industrial storage systems—we're constantly told the same story:

"We're storing energy for later use."

This story has shaped the entire energy economy. It has defined how we build infrastructure, how we plan mobility, and how we imagine the future of sustainability.

But what if it's wrong?

What if, all this time, we've misunderstood what a battery actually is?

Because here's the uncomfortable truth from the perspective of Universal Energy Dynamics (UED):

A battery doesn't store energy. It stores unresolved discomfort in the ripple field.

The Tank Metaphor Has Misled Us

We've been conditioned to treat a battery like a fuel tank:

- You "fill" it with energy.

- It "holds" energy inside.
- Then you "drain" it when you need power.

It's a convenient metaphor—but completely misleading.

There are no particles of energy inside a battery. There's no glowing orb waiting to be released. There's no bucket of electricity sloshing around.

When you open a battery, you don't find energy. You find chemicals, ions, potential differences, and most importantly—tension.

Not tension between materials. Tension between field configurations.

So What Is Inside a Charged Battery?

A charged battery is a spatially structured imbalance.

Its internal chemistry has been rearranged—electrons displaced, ions stacked, interfaces distorted—in a way that holds the field in a non-equilibrium state.

This state is not *energy-rich*. It's alignment-poor.

It's not a container—it's a contradiction. And the field doesn't like contradictions.

So the moment you connect a circuit, what happens?

The field tries to resolve the misalignment. That attempt to restore balance is what we observe as current.

The Premise of This Chapter

This chapter invites a fundamental shift in how we think about storage.

We're not saving energy. We're postponing field realignment.

A battery is not a vault. It's a prolonged disturbance.

It doesn't give us power. It gives us a situation the universe wants to fix.

And every time the universe tries to fix it—light glows, motors spin, circuits live.

That's not extraction. That's permission.

Key Insight

"There is no energy inside. Only structured field discomfort, waiting for an opportunity to dissolve."

Next, we'll explore what a battery actually is—not chemically, but ripple-logically.

8.1 What Is a Battery, Really?

Strip away the marketing gloss, the chemical jargon, and the buzzwords like "energy-dense" or "high-capacity," and ask yourself: What is a battery, fundamentally?

Not metaphorically. Not functionally. But structurally—from the field's point of view.

The answer under Universal Energy Dynamics (UED) is simple and provocative:

A battery is not a container of energy. It is a geometric field imbalance—a carefully sustained misalignment in the ripple configuration of matter and space.

Batteries Don't Store Energy—They Hold Tension

In mainstream science, a battery is described as a device that stores and releases chemical energy through redox reactions. Electrons move through a circuit, and ions shuffle between electrodes via an electrolyte.

But this only describes the mechanism, not the meaning.

UED offers a deeper perspective:

- During charging, the battery is forced into an unnatural field configuration.
- The chemistry becomes a carrier of that distortion—but the real strain is in the surrounding Structured Vacuum Energy (SVE).
- The two terminals represent zones of different ripple

alignment, separated by an internal struggle.

The battery isn't full of energy. It's full of unresolved alignment—held like a stretched rubber band that hasn't snapped back.

Electrochemical Reactions = Ripple Stressors

What we call oxidation and reduction are not just electron trades. They are ripple reformatting events.

Each movement of ion or electron shifts the local ripple geometry of the atomic lattice. These shifts aren't trivial—they generate tension footprints in the vacuum field.

As ions accumulate on one side and vacate another, they create a polarized ripple gradient—a misalignment that the field inherently wants to resolve.

Think of it as a disagreement between two zones of reality, held apart by chemical bouncers.

The battery's job? Hold that disagreement stable until someone opens the door.

The ATT Perspective: Temporarily Suspended Alignment

In UED terms, the battery during charge is in an active ATT state:

- A – misalignment exists between the poles.
- T – the imbalance is sustained over time.
- T – the field tension is elevated and held in check.

This ATT condition is not stable—it's suspended.

The moment a circuit is completed, ATT collapses, and the field seeks a new balance by moving ripple tension through the wire geometry.

We see this as current, voltage, and discharge.

But what's really happening is this:

The field is breathing out a discomfort it's been holding in.

A Battery Is a Misalignment Device

Let's define it formally:

A battery is a ripple-structured device that maintains electrochemical misalignment across spatial terminals, holding the field in a deliberate non-equilibrium state, waiting for a chance to resolve.

That's not energy. That's a ripple contradiction—neatly packaged for later correction.

And the field is always listening, always ready to heal—if we give it a path.

Closing Thought

"A battery doesn't store power. It stores the privilege of future realignment."

What we've been calling energy is just the ripple tension that's been politely waiting to be released.

In the next section, we'll go deeper: What really happens during charging and discharging? And why is storage not containment—but delayed resolution?

8.2 Storage Is Not Containment—It's Delayed Resolution

We have been taught to picture energy storage like stocking up grain in a warehouse or water in a tank. The language is everywhere:

- "Store energy."
- "Save it for later."
- "Release power when needed."

This metaphor is so embedded in our mindset that we barely question it. But it only makes sense if you believe energy is a thing—a transferable substance that moves in and out like liquid.

Universal Energy Dynamics (UED) shows that this is not only incorrect—it's dangerously misleading.

Energy is not stored. It is postponed. And storage is not containment—it's discomfort deferred.

What Happens During Charging?

In mainstream terms:

- You plug in a device.
- Current flows into the battery.
- Electrons are pushed into chemical structures.
- Energy is "stored."

But in UED terms, here's what really happens:

- Charging imposes a ripple distortion across the battery's internal geometry.
- This creates a ripple gradient—a misalignment between the two terminals, sustained by the strained electrochemical lattice.
- The longer the charge is held, the more tension builds in the field geometry surrounding the battery.

This is not a tank being filled. It's a spring being stretched, or better—a ripple pattern being held in suspense, against the vacuum's natural inclination to restore balance.

The field is not energized. It's uncomfortable. And it remembers.

What Happens During Discharge?

Discharging a battery is not about releasing a substance. Nothing flows out. Nothing escapes.

Instead, here's what occurs under UED:

- A path to resolution is provided—via a completed circuit.
- The ripple misalignment begins to collapse.

- Field tension moves—not as particles, but as reconfiguration.
- What we perceive as electric current is just the field snapping back into alignment.

Light glows, motors spin, heat emerges—but not because energy "left" the battery.

It's because misalignment resolved, and the ripple field rebalanced itself through the geometry of the connected device.

Discharge = Field relief. Not release of a stored thing, but completion of an uncomfortable pattern.

Spring Analogy, Upgraded

We often compare batteries to springs—charge compresses the spring, discharge lets it expand.

But UED gives this analogy much sharper teeth:

- A battery is not a spring inside itself.
- It's a field spring—tension not just held in matter, but rippling through the surrounding vacuum.
- The device doesn't contain the potential. It anchors it in field geometry.

The "stretch" exists in the vacuum's structure, not the battery's box.

Why This Matters

Why reframe something that seems to work fine?

Because the misunderstanding of storage as containment has led to:

- Inefficient designs based on the idea of filling and draining.
- Dangerous assumptions about limitless scaling ("just build bigger batteries").
- Ignorance of ripple side effects—leakage, overheating,

chaotic field behavior.
- Disconnection from the field's intelligence—treating it like a warehouse instead of a living system.

If you think you're managing a tank, you'll keep trying to seal it tighter. But if you realize you're holding the field in suspense, you'll focus on how gently and coherently that tension can be created and resolved.

Key Insight

"Charging doesn't add energy. It introduces a ripple contradiction. Discharge is not energy release—it's field forgiveness."

8.3 Capacitors vs Batteries – Two Strain Logics

At first glance, capacitors and batteries seem to do the same job: They store energy for later use. But their behavior, performance, and applications are vastly different.

- Capacitors charge and discharge almost instantly.
- Batteries charge and discharge over time.
- Capacitors are used in circuits for timing, filtering, and buffering.
- Batteries are used for sustained power delivery.

Mainstream science explains this by saying:

Capacitors store energy in an electric field. Batteries store energy in chemical bonds.

But this view is still trapped in the myth of storage—treating energy like a substance to be poured in and poured out.

Universal Energy Dynamics (UED) sees it differently.

Neither capacitors nor batteries store energy. They both hold misalignments—just in different ways, and over different timeframes.

Let's unpack their distinct strain logics.

Capacitor: Short-Term Field Tension

A capacitor consists of two conductive plates separated by a dielectric (non-conductive) material.

- When charged, electrons accumulate on one plate and are removed from the other.
- This creates a ripple imbalance across the plates—a pure electrostatic misalignment.
- No chemistry, no moving parts, no ions. Just geometric ripple tension held in place by insulation.

In UED terms:

- The capacitor pinches the field between its plates, holding it off balance.
- The Structured Vacuum Energy (SVE) tries to resolve the difference—but is prevented from doing so by the dielectric barrier.
- As soon as a path is provided (i.e., discharge), the field snaps back into alignment, releasing the tension as a brief, sharp current pulse.

A capacitor is like a field spring loaded and released in milliseconds.

It excels at fast tension exchange, but cannot hold that tension for long.

Battery: Long-Term Ripple Misalignment

A battery, on the other hand, is a chemically sustained ripple disruption.

- It uses redox reactions (electron exchanges) to build a ripple gradient between two terminals.
- This gradient is not merely electrostatic—it is entangled with chemical transformations in the electrodes and electrolyte.
- The resulting misalignment is embedded in the atomic structure of the materials.

In UED terms:

- A battery doesn't just distort the field—it reconfigures the material geometry in a way that locks in the misalignment.
- The field now has to negotiate through chemistry to realign—making the discharge process slower, more controlled, and longer lasting.

A battery is like a field contradiction woven into matter—capable of persisting for hours, days, or years.

It trades speed for endurance.

Two Logics, One Principle: Delayed Realignment

Though capacitors and batteries differ in mechanism and timescale, they share a deeper truth:

Neither holds energy. Both hold delayed alignment events.

- The capacitor is a fast discomfort device. → It charges quickly, discharges explosively, and is done.
- The battery is a slow discomfort device. → It takes time to disturb the field, and time to release that strain.

But in both cases, what's released is not a substance—it's the field's effort to restore itself.

They are not energy banks. They are scheduled ripple responses.

Why This Matters in Design

If you see both as "energy storage" devices, you'll:

- Compare them by capacity, voltage, and output curves.
- Focus on "how much" they hold.

But under UED:

- You focus on how they disturb the field.
- You design for coherence of misalignment, smoothness of resolution, and minimization of unintended turbulence.

This opens up a new era of field-informed design:

- Materials tuned for ripple alignment.
- Architectures that delay resolution with minimal strain.
- Devices that don't fight the field—but flow with its natural response cycles.

Closing Insight

"A capacitor is a ripple breath held briefly. A battery is a field story paused in chemistry."

"Neither stores energy. They simply wait for the universe to correct what they disturbed."

Next, we'll go deeper into the fallacy of "energy inside" and what mass, heat, and strain really represent inside so-called storage devices.

8.4 There Is No "Energy Inside"

We open our smartphones and see a little icon that tells us we're at 83% battery. We plug in power banks and imagine them filled with "juice." We refer to batteries as "charged," "full," "dead," or "empty."

The entire language of modern technology rests on this unspoken assumption:

That energy is inside the battery.

But Universal Energy Dynamics (UED) reveals a more accurate—and more profound—understanding:

There is no energy inside. There is only a configuration that the universe is waiting to correct.

A battery is not a vault. It's not a fuel tank. It's not a capsule of stored energy.

It's a held contradiction—a persistent discomfort in the ripple alignment of the structured vacuum field. And every time you draw "power" from it, all you're doing is giving the universe permission to breathe out that discomfort.

What the Battery Actually Contains

From a classical view:

- The battery holds "chemical energy" in the bonds between atoms.
- During use, these bonds break or shift, releasing electrons that move through a circuit.

This is not wrong in a mechanical sense—but it's misleading in a fundamental one.

From the UED perspective:

- Those chemical bonds are not energy stores. They are ripple alignments—configurations of matter that create tension in the surrounding field.
- Charging forces those alignments into an unnatural ripple pattern—one that does not sit well with the field.
- The battery holds an uncomfortable situation that the vacuum wants to resolve.

It's not what the battery has that matters. It's what the universe feels about its state.

Why Charging and Discharging Generate Heat

We often think of heat during charging and discharging as "loss"—as if the battery is inefficient because some energy "leaks out" as warmth.

But under UED, heat is not leakage—it's evidence of ripple turbulence.

- When a battery is charged too fast, the ripple misalignment is forced too quickly.
- When discharged too fast, the field is forced to reconfigure rapidly, creating chaotic ripple realignments.

Heat is not excess—it's strain.

- It reveals how uncomfortably the field is being disturbed.
- It shows how far from smooth alignment the process has become.

That warmth is the field saying,
"This resolution is messy. I'm trying to catch up."

Why Batteries "Die"

Eventually, every battery stops working. We say it's "dead." We assume it has "run out of energy."

But that's not what's happening.

A battery dies not because it ran out of something, but because its ability to hold misalignment has degraded.

Over time:

- The internal structure corrodes.
- The ripple geometry becomes inconsistent.
- The chemical pathways no longer maintain a clean polarization.

The battery is no longer able to create a tension the field cares about.

It's like trying to stretch a spring that has lost its stiffness. The ripple field no longer sees a meaningful contradiction—and thus, it stops responding.

There's no release because there's nothing left to resolve.

No Fuel. No Flow. Just Field Conditions.

When you realize this, the paradigm shifts:

- "Charge" is no longer a thing—it's a condition.
- "Full" is not a quantity—it's a tension state.
- "Power output" is not substance leaving—it's field realignment happening.

Batteries aren't engines. They're ripple invitations—begging the universe to come clean up the mess they've been holding.

And when that invitation expires? The show is over.

Closing Insight

"A battery doesn't run out of energy. It runs out of the ability to disturb the field in a meaningful way."

"The field never stopped responding. The battery just forgot how to ask."

Next, we'll explore how the efficiency of storage systems should be measured—not by energy density, but by how gracefully they hold and release ripple strain.

8.5 Why Efficiency Depends on How You Store Discomfort

In classical engineering, a battery is considered efficient when:

- It stores a large amount of energy per unit volume (high energy density),

- Delivers most of that energy to the load (low internal loss),
- And maintains performance over many cycles (good lifespan).

But this entire view is built on the flawed assumption that energy is a stored thing—something you load in, hold, and extract.

Universal Energy Dynamics (UED) turns this paradigm on its head.

In UED, efficiency is not about *how much* you store, It's about *how well* you hold misalignment—and how gently you let the field resolve it.

In other words:

Efficiency = the elegance of discomfort management.

Let's explore what this really means.

It's Not the Density—It's the Discipline

Modern battery tech races to pack more energy into smaller volumes: lithium-ion, graphene, solid-state, sodium-air, and more.

But UED asks:

- What *pattern* of misalignment are you imposing on the field?
- How *coherent* is the ripple strain across the battery's internal geometry?
- Can the field resolve it smoothly, or does it have to struggle?

Ripple-aligned storage means:

- The chemical structure supports a stable, symmetric misalignment.
- The ripple tension is clear, localized, and easy to guide.
- When you connect a circuit, the field resolves smoothly, with minimal turbulence.

Result?

- Less heat.
- Longer lifespan.
- Stable voltage curves.
- Near-lossless operation.

Ripple-chaotic storage, on the other hand:

- Holds misalignment in confused, unstable, or distorted ripple geometries.
- May use complex chemistries or poor lattice structures that strain the field unevenly.
- Field response becomes turbulent, noisy, or partial.

Result?

- Heat generation.
- Voltage sag or spikes.
- Internal degradation.
- Fast capacity fade.

Heat, Leakage, and Decay = Misalignment Noise

These aren't inefficiencies—they're symptoms of bad ripple hygiene.

- Heat during charging/discharging? That's the field struggling with incoherent correction.
- Self-discharge? That's the field passively trying to resolve disorder even when you're not using the battery.
- Capacity degradation? That's internal geometry losing the ability to hold a clean misalignment.

Just as a singer loses clarity when off-key, a battery loses efficiency when its internal ripple isn't in harmony.

Efficient storage isn't dense—it's graceful.

Designing for Ripple-Coherent Storage

Under UED, the future of energy storage doesn't lie in:

- Bigger voltages,
- Higher pressures,
- Or more exotic materials.

It lies in alignment literacy:

- Understanding how ripple strain propagates through electrochemical geometries.
- Building materials that welcome and hold tension elegantly.
- Designing architectures that allow clean, symmetric resolution without chaos.

Imagine batteries that:

- Never heat up.
- Never degrade.
- Release power as a field poem, not a chemical fight.

This isn't science fiction. It's what happens when we stop fighting the field and start speaking its language.

Closing Insight

"Efficiency is not about how much discomfort you can hold. It's about how little pain the field feels while helping you resolve it."

"A perfect battery doesn't store more—it asks less, and aligns better."

In the final section, we'll revisit the battery not as a power source, but as a handshake with the universe, delayed until we allow it to complete.

8.7 Final Insight – A Battery Is Not a Box. It's a Bargain

We've spent decades trying to "put more energy in the box." We've built lithium cells, stacked electrode layers, and engineered nanostructures—all to store more.

We imagine that carrying a battery means carrying power. But under the lens of Universal Energy Dynamics (UED), a radical truth emerges:

You're not carrying energy. You're carrying a promise.

That battery in your phone, car, or satellite is not holding power. It's holding a ripple misalignment—a field disturbance the universe has *agreed to tolerate* for a while... until you let it go.

Storage = Postponed Participation

When you "charge" a battery, the universe doesn't give you energy. You simply disturb the field and convince it to wait before resolving that discomfort.

That's not storage. That's a deferred alignment event.

The battery is not a box of fuel. It's a time-delayed field contradiction—held, not hoarded.

And when you use the battery? You're not "releasing energy." You're allowing the universe to correct the discomfort you once introduced.

It's not a flow. It's a fulfillment.

Design Shifts: From Material to Meaning

Mainstream science chases new chemicals to hold "more energy":

- Denser lithium packs.
- Graphene supercaps.
- Solid-state electrochemistry.
- Metal-air experimental cells.

But under UED, the chemical isn't the problem. The geometry of the misalignment is.

The field doesn't care if your battery is made of lithium, sodium, or wishful thinking. It only responds to how cleanly the discomfort is configured.

Thus, real breakthroughs won't come from better materials. They'll come from better field conversations.

We don't need to squeeze more volts into tighter spaces. We need to design devices that negotiate ripple misalignment with minimal noise and maximal elegance.

The future of storage is not brute density. It's alignment fluency.

Bigger Isn't Better. Smarter Is.

For too long, "better battery" meant:

- Bigger capacity,
- Longer runtime,
- Faster charging.

But every attempt to push harder has led to:

- More heat,
- Faster degradation,
- Greater environmental cost.

Why?

Because we've been trying to bully the field, not bargain with it.

The field isn't impressed by scale. It responds to precision—to how intelligently you introduce, hold, and release strain.

In that light:

- A small, cleanly-aligned capacitor can outperform a large, chaotic battery in certain contexts.
- A nano-engineered ripple lens may replace entire voltage

stacks.

- "Storage" may someday mean modulating the vacuum directly—not storing anything at all.

Final Reflection

"You don't carry energy in your pocket. You carry a delay the universe agreed to resolve later."

"Every battery is a ripple agreement. Every discharge is a fulfillment of that pact."

Storage is not about possession—it's about timing. And timing, in a universe made of patterns, is everything.

In the next chapter, we'll extend this ripple logic to the global scale—revealing how power grids are not pipelines, but massive synchronized ripple negotiations where nothing truly flows.

Chapter 9: Power Grids and the Illusion of Flow

Every night, cities light up. Skyscrapers glow. Trains glide. Phones charge. Satellites sync. And behind it all, one idea hums silently in the background:

That electricity is flowing.

From hydro plants to households, from coal stations to computer servers, from solar farms to street lamps—we believe there is a substance moving across vast distances through wires, poles, and transformers.

We visualize:

- Dams "sending power" to villages.
- Power plants "supplying" cities.
- Grids "delivering" electricity through metal veins.

It's one of the most convincing illusions in the modern world.

But it's still an illusion.

Because in truth— Nothing flows.

Not electrons. Not energy. Not charge.

What really moves across the grid is ripple tension—a disturbance in the field, not a material stream.

And in this chapter, we will reveal the real nature of electrical grids under Universal Energy Dynamics (UED):

Power isn't transmitted. Ripple discomfort is distributed. And what we call electricity is simply structured misalignment seeking resolution.

The Problem of Language and Visualization

We've been taught to visualize electric current like water in a pipe:

- It flows from high voltage to low.
- It carries energy.
- The bigger the pipe (wire), the more energy you can push through.

But this metaphor is deeply flawed.

It assumes:

- That something tangible is moving inside wires.
- That the energy "starts" in one place and "arrives" somewhere else.
- That the wire is a transport vessel, not an interaction zone.

This visualization worked for the industrial age—but it's outdated.

The grid isn't a system of pipelines. It's a network of ripple-responsive geometries.

And once we see this, everything from transformers to voltage drops to frequency synchronization begins to make more sense—not as anomalies, but as field behavior.

The Grid Is Not a Highway. It's a Ripple Web.

When a hydropower plant in Bhutan operates, and a house in Delhi turns on a light, we say, "the power was transmitted."

But UED tells us:

- The field geometry between the generator and the load

forms a coherent ripple zone.
- The generator initiates a structured disturbance (an oscillating tension).
- That disturbance propagates through field alignment, not through material flow.
- The bulb doesn't receive energy—it becomes the point of ripple resolution.

Just like sound is not air moving from a speaker to your ear—Electricity is not charge moving from a dam to your city.

It's the field vibrating, and the structure responding.

The Premise of This Chapter

In this chapter, we will dissolve the illusion of electric flow:

- By breaking down the misunderstood nature of alternating current (AC),
- By reframing wires and transformers as ripple conduits—not power pipes,
- And by showing how the entire grid operates as a synchronized ripple negotiation, not a delivery system.

We will also challenge:

- The myth of electron drift as the mechanism of power transfer,
- The belief that "energy travels" through conductors,
- And the fantasy that a city can "import electricity" from another state.

There is no import. There is no movement. There is only shared tension across a coherent field.

Closing Line for the Opening Hook

"You're not receiving electricity. You're participating in the field's discomfort—until it resolves itself in the geometry of your device."

Let's now decode how this "participation" actually works—starting with the alternating current that powers most of the planet.

9.1 How AC Really Works

Every home, city, and industrial complex today runs on Alternating Current (AC).

You flip a switch in Mumbai, and a ceiling fan begins spinning. You start a laptop in Berlin, and it draws current from the wall. All of this happens thanks to the miracle of AC.

But what is really happening in those wires?

The mainstream explanation is familiar:

- Electrons are pushed and pulled by a sinusoidal voltage.
- They move back and forth in the conductor.
- This "oscillating flow" delivers energy to the load.

It sounds logical. Even elegant. But it's not what's actually happening.

Under Universal Energy Dynamics (UED), Alternating Current is not a flow of electrons. It is a periodic ripple disturbance—a tension wave in the Structured Vacuum Energy (SVE), oscillating in rhythm with the source geometry.

Let's decode it.

No Back-and-Forth Electron Dance

Textbooks often describe AC current as electrons vibrating in place—moving a fraction of a millimeter one way, then reversing direction, 50 or 60 times per second.

But even this model is misleading.

Electrons, under UED, are not particles, but localized ripple knots—stable field distortions. They do not shuttle back and forth like marbles in a tube.

There is no need for charge carriers to "move" to deliver energy. In fact, nothing needs to move at all.

What matters is the oscillation of ripple tension—how the field geometry is being rhythmically disturbed and reconfigured.

AC is not a back-and-forth flow. It's a pulsed discomfort, imposed and resolved in time.

The True Nature of Alternating Current: A Ripple Tension Wave

Think of a still lake. Now imagine tapping it rhythmically at one end. You don't move water from one side to the other. You create waves—disturbances that propagate through the field.

Similarly, in AC:

- The generator oscillates its geometry (typically a coil in a magnetic field),
- This creates a cyclical ripple misalignment in the field around it,
- That disturbance propagates through the grid's aligned geometry—from generator to devices.

There is no need for mass or charge to travel. What travels is the pattern—the structured discomfort.

And this pattern is locked to frequency:

- 50 Hz in India and Europe,
- 60 Hz in the U.S.

This frequency is not a byproduct—it is the heartbeat of ripple misalignment.

Field Reconfiguration, Not Electron Transport

When you plug in a device:

- It doesn't "receive" electrons from far away.
- It becomes the end point where the ripple tension resolves itself locally.
- The filament glows, the motor turns, or the speaker vibrates—not because something arrived, but because the field rearranged itself in that geometry.

In UED terms:

AC current is a structured field oscillation, And electrical devices are zones of ripple translation—not endpoints of delivery.

Every Cycle Is a Negotiated Rebalance

Each cycle of AC is a complete event:

- Tension is imposed → misalignment propagates → resolution begins → reversal.
- This bidirectional rhythm ensures that the system stays in dynamic balance.

This is why AC is more efficient for long-distance transmission:

- The field is never fully strained in one direction.
- It continuously re-aligns itself, reducing ripple resistance and enabling large-scale synchronization.

In a way, AC doesn't fight the field—it dances with it.

Closing Insight

"Alternating current is not a current at all. It is a rhythmic ripple correction—a universal exhale and inhale through structured geometry."

Next, we'll examine what transmission lines really are—not pipes, but field resonators tuned to carry ripple alignment, not energy.

9.2 Transmission Lines as Field Resonators

We often imagine power lines as energy highways—huge, humming cables transporting electric charge across landscapes, delivering energy like a convoy of trucks moving fuel.

This mental image is everywhere:

- Towers stretch over mountains.
- Cables buzz in the monsoon.
- News headlines speak of "power import" and "energy evacuation."

But here's the truth from Universal Energy Dynamics (UED):

Transmission lines do not carry energy. They conduct ripple tension.

They are not pipes, but field resonators—geometrically structured to allow ripple misalignment to propagate cleanly from one region to another.

Let's strip away the myth and look at what transmission lines actually do.

Voltage = Ripple Tension

In classical physics, voltage is defined as electric potential difference—how much "push" a unit charge would experience moving from one point to another.

But under UED, voltage is not a potential. It is a field condition.

Voltage is a measure of ripple tension across a spatial geometry.

The higher the voltage:

- The more strained the ripple field between two points.

- The more urgency the field feels to realign.

High-voltage lines don't "push" more charge. They introduce a stronger field discomfort that can propagate over longer distances with minimal ripple distortion.

That's why long-distance lines operate at 400kV, 765kV, or even more—not to carry more electrons, but to support sharper, cleaner ripple gradients.

Frequency = Rhythm of Misalignment

Every alternating current (AC) system oscillates at a certain frequency:

- 50 Hz in India, Europe, and Asia.
- 60 Hz in the U.S. and parts of Japan.

But why?

Because frequency, in UED terms, is not just a convention. It is the rhythm at which the field is cyclically disturbed.

Frequency is the pulse rate of misalignment cycles across the grid.

Transmission lines are designed to resonate at this frequency. Like strings on a violin, they are tuned to pass the ripple tension with minimum loss and maximum coherence.

Change the frequency—and the entire field response shifts.

Power = Rate of Discomfort Negotiation

In classical terms:

- Power = voltage × current.
- It represents "rate of energy transfer."

But what's *actually* transferring?

Under UED, power is not the movement of anything.

It is the rate at which ripple discomfort is being negotiated across the field.

- High power does not mean more energy is being delivered.
- It means the field is being disturbed and resolved more rapidly, more intensely.

Think of it like a conversation:

- Low power = quiet discussion.
- High power = heated debate.
- But in both, nothing is being passed—just realignment happening faster.

What Are Transmission Lines, Really?

Let's redefine them:

Transmission lines are spatial geometries engineered to maintain coherent ripple tension propagation over distance, with minimal dissipation, tuned to a resonant rhythm.

They are field instruments, not wires.

Their success lies not in how much they carry, but in how little they distort the ripple as it moves.

Every tower, every insulator, every cable gauge exists to maintain:

- Clean field alignment,
- Stable ripple rhythm,
- And minimal side-turbulence (which we perceive as losses).

Rethinking Losses

Transmission losses aren't because electrons get tired. They are due to:

- Imperfect alignment,

- Ripple turbulence,
- Thermal agitation where field strain gets trapped.

If the geometry is coherent, losses drop. If not, the field throws off heat, noise, and harmonics.

So, the future of transmission isn't thicker wires. It's better field geometry—smarter resonators, not bigger carriers.

Closing Insight

"A power line doesn't carry anything. It tunes the vacuum to carry the discomfort forward."

"Voltage is tension. Frequency is rhythm. Power is the ripple's urgency to return to stillness."

Next, we'll dismantle one of the most persistent myths in electrical science: the Electron Drift Fallacy.

9.4 The Electron Drift Fallacy

One of the most peculiar contradictions in standard electrical theory is this:

"Electricity moves at nearly the speed of light... but electrons drift at less than a millimeter per second."

Wait—what?

How does the bulb light up instantly when you flip the switch, even though the electrons are crawling like snails through the wire?

This inconsistency has puzzled students, confused engineers, and been swept under the rug by convenient metaphors like "energy moves, not matter."

But here's the real problem:

The entire concept of electron drift as a mechanism of energy transfer is flawed.

In fact, under Universal Energy Dynamics (UED), the electron itself is misunderstood.

It is not a particle traveling through the conductor. It is a localized ripple knot—a stable distortion in the structured vacuum field.

And it doesn't need to move at all.

The illusion of "current" is not due to drifting charges. It is the ripple field itself realigning—not by pushing anything, but by coherent tension adjustment through space.

What the Textbooks Say: The Drift Story

In conventional physics:

- Electrons in a wire are described as forming a "sea of mobile charge carriers."
- When voltage is applied, these electrons drift slowly in one direction (in DC), or vibrate back and forth (in AC).
- The energy is said to move fast, but the electrons... barely crawl.

To explain the fast lighting of a bulb, we're told that:

"It's like a pipe filled with marbles. Push one in, one pops out at the other end."

But this is just a metaphor—and a lazy one at that.

It presumes:

- Charge behaves like matter.
- Flow is necessary to explain transfer.
- Electrons are the energy carriers.

All false under UED.

UED View: Electrons Are Anchored Ripples

From the UED perspective:

- Electrons are not particles. They are field distortions—ripple geometries that maintain a local

imbalance.
- In a conductor, electrons are part of the lattice geometry—not free-moving billiard balls.
- What we perceive as "current" is not their movement, but the ripple reorientation happening across the structure.

Imagine a rope held between two people. You shake one end, and the wave reaches the other almost instantly—but nothing traveled.

That's how current works.

The field moves. The pattern shifts. The "carriers" remain mostly in place.

Why the Drift Fallacy Persists

The electron drift idea persists because:

- It matches our outdated flow-based models.
- It gives a false sense of cause-effect ("movement = work").
- It's mathematically easy to model in circuits.

But it fails to explain:

- Instantaneous signal propagation.
- The mismatch between motion speed and response.
- Why thermal effects occur even when no net drift happens (as in AC).

And most importantly, it hides the truth:

The wire is not a conduit. It's a field zone.

Ripple Reorientation = True Cause of Current

So what really causes electric current?

Ripple reorientation.

When a voltage is applied:

- The ripple tension in the field geometry changes.

- Each part of the lattice responds by realigning its local field.
- This realignment cascades across the structure—fast, coherent, and near-instant.

No mass is transported. No particles need to move.

Just field discomfort resolving through the path provided.

It's like tilting a table and seeing marbles roll—not because marbles push each other, but because the surface geometry changed, and the marbles responded.

Closing Insight

"Electrons don't drift. The illusion of flow is the consequence of ripple choreography, not particle transport."

"Electricity is not a substance in motion. It is a pattern in transformation—field discomfort cascading through geometry."

Next, we explore the consequence of this truth at the grid level: How devices across the world become local resolution points, not receivers of power.

9.5 Load = Local Resolution Point

We often hear things like:

- "Bhutan exports hydroelectricity to India."
- "The power generated at the dam in Tehri is being supplied to Delhi."
- "Energy flows from generation to consumption."

These statements are repeated in engineering, policymaking, and everyday conversations. But they rest on a deeply flawed premise:

That energy travels.

From the UED perspective, this is not what happens.

A light bulb in Delhi does not receive energy that traveled all the way from a distant hydroelectric dam. It experiences field realignment—a local event triggered by a ripple condition initiated and maintained by the geometry of the entire grid.

In simpler terms:

A load is not a destination. It is a location where ripple discomfort resolves.

The Myth of Point-to-Point Delivery

In the traditional view:

- A generator creates energy.
- This energy is transmitted across wires.
- A load consumes it at the other end.

But this visualizes the grid like a pipeline—a point-to-point delivery system where power has a starting point and an endpoint.

This works for physical substances like water or oil. But electricity is not a substance—it is field behavior.

You cannot "send" a field alignment the way you send a parcel.

The reality is more subtle and more profound.

Ripple Collapse Is Geometric, Not Geographic

The moment a switch is flipped in Delhi:

- A path of potential field resolution is established through the grid.
- The structured vacuum, through conductive material, detects a misalignment opportunity.
- The ripple tension—already present and oscillating due to the generator in Bhutan—finds its collapse point locally.

No electron travels from Bhutan to Delhi. No energy is transmitted like a package.

What happens is this:

The entire grid functions as one massive ripple chamber, And the geometry of devices, not their distance from the source, determines where the ripple collapses.

Load = Completion Point of Ripple Handshake

From a UED perspective:

- A generator imposes rhythmic discomfort.
- A transmission network maintains coherent field strain.
- A load—like a light bulb or motor—is a place where the field finds an opportunity to resolve part of that strain.

Thus, a load is not a consumer. It is a completion point—a ripple handshake terminal.

The power plant doesn't send energy to the load. It sends tension into the grid, and the load simply receives permission to act as a resolution site.

There is no transport. Only coordinated field response.

Implications for Grid Thinking

This reframing changes how we think about:

- Grid balance: It's not about load sharing, but about ripple coherence.
- Power cuts: Not due to lack of delivery, but due to disruption in field tension negotiation.
- Smart grids: Should not just manage supply-demand statistics, but map ripple dynamics across time and topology.

A city doesn't receive power. It offers locations for ripple comfort—and the universe responds accordingly.

Closing Insight

"Power doesn't arrive. It collapses locally."

"The bulb doesn't light because something was delivered. It lights because the universe saw a path to resolve its tension—and took it."

Next, we'll expand this ripple logic to the entire grid and show why power grids aren't delivery systems at all, but coherent ripple networks designed for synchronized discomfort.

9.6 Grids as Ripple Systems, Not Delivery Networks

When we think of the power grid, we tend to visualize it as a giant supply chain—a vast network where "energy" is generated at one end, transported through high-voltage lines, transformed, distributed, and finally consumed at the other end.

This visualization is familiar. It seems logical. It's also wrong.

The grid is not a delivery system. It is a ripple tension network—a single, dynamic field of structured discomfort and coordinated realignment.

Under Universal Energy Dynamics (UED), the entire grid behaves not as a set of energy pipes, but as a coherent ripple chamber—a web of geometry tuned to conduct, distribute, and resolve field tension across space.

The Grid as a Unified Field Zone

At any moment, the grid spans:

- Thousands of kilometers,
- Multiple generation sources (hydro, solar, thermal, nuclear),
- Millions of devices drawing power (loads),
- Billions of ripple reconfigurations per second.

Yet it behaves as one field.

Why?

Because the entire system is electromagnetically synchronized:

- At a common frequency (e.g., 50 Hz in India).
- With a shared ripple rhythm.
- Tuned for coherent ripple propagation through conductive materials and field-aligned geometries.

There are not many power grids. There is one ripple system partitioned by policy—not physics.

Generators = Controlled Discomfort Sources

In classical models, a generator "produces energy." But in UED, a generator's real role is:

To introduce rhythmic discomfort into the field.

By rotating a coil in a magnetic field (or vice versa), generators:

- Disturb the local ripple equilibrium,
- Modulate ripple tension at a defined frequency and amplitude,
- Feed this oscillating field misalignment into the grid.

Each generator becomes a tension pump—not delivering anything, but agitating the field rhythmically and geometrically.

Different generators can be:

- Coherent (adding aligned ripples), or
- Incoherent (causing destructive interference).

That's why synchronization across the grid is vital—not to coordinate deliveries, but to preserve ripple harmony.

Loads = Invitations to Resolve Tension

Loads (lights, motors, heaters, chargers) do not "consume energy" in the literal sense. They provide localized field geometries that invite ripple resolution.

When a load is connected:

- It introduces a ripple pathway where the field can collapse its existing tension.
- This completes the ripple handshake initiated by the generator.

Thus, the load is not a drain. It is a resolution node—a relaxation portal in the ripple network.

Multiple loads across the grid invite resolution at different geometries and scales, enabling the grid to breathe tension in and out continuously.

Synchronized Imbalance: The Grid's True Operation

The grid does not run on energy flow. It runs on managed misalignment.

Here's how it works:

- Generators introduce oscillating discomfort.
- Transmission lines maintain structured ripple resonance.
- Loads allow partial resolution.
- The system never fully stabilizes—because perpetual imbalance is necessary for function.

This is the paradox:

The grid is alive because it never fully calms down.

It is a choreographed chaos—a tension field pulsing in rhythm, collapsing and reviving misalignment with every cycle.

That's why sudden load drops or spikes cause instability—not because of material shortages, but because the ripple negotiation was disrupted.

Closing Insight

"The grid doesn't deliver electricity. It distributes discomfort—and guides its resolution."

"Generators shake the field. Loads calm it down. Transmission lines tune the dance."

In the next section, we'll explode the popular notion that electricity "travels thousands of kilometers"—and show why long-distance transmission is not transport, but resonance extension.

9.7 The Myth of Long-Distance Energy Transfer

We are told that electricity generated at a hydropower plant in the mountains can be sent over 1,000 kilometers to light up homes in faraway cities.

Maps proudly show transmission lines stretching across states, countries, ev.en continents—"delivering" electricity as though it were milk in a pipeline.

This visual makes infrastructure seem grand and energy seem mobile.

But what if nothing actually moves?

What if this entire concept of sending energy across distances is just an elegant illusion—a misinterpretation of what's really happening?

Under Universal Energy Dynamics (UED), that's exactly the case.

Energy is not transmitted. It is reconfigured locally—wherever ripple tension finds a compatible resolution geometry.

Let's unravel the myth.

There Is No Cross-Country Energy Flow

The mainstream explanation says:

- Power is generated at a dam in Bhutan.
- It travels via transmission lines to a city like Delhi.
- It is delivered to homes, factories, and grids.

But what does this actually mean?

No single electron makes this journey. No packet of energy is boxed, shipped, or passed along the wire like a baton.

What actually happens is this:

- The field tension created by the generator establishes a ripple condition across the grid.
- That ripple interacts with the structured geometry of the transmission line.
- Devices in Delhi don't receive that energy—they resolve their own local ripple strain, in rhythm with the tension cycle that originated elsewhere.

The generator doesn't *send* energy. It offers a field vibration. The devices respond if they are tuned to it.

Just like a guitar string can vibrate in harmony with a piano string across the room—without touching—so too does the light bulb in Delhi resonate with the tension geometry sourced from Bhutan.

No travel. Just ripple alignment.

Transmission Lines Are Alignment Pathways

High-voltage transmission lines are not "energy highways" that carry something.

They are:

- Ripple-capable structures, designed to conduct field misalignment coherently.
- Resonance paths, allowing ripple tension to propagate without significant distortion.
- Alignment bridges, not transport channels.

Electricity doesn't go *through* a transmission line. The ripple *passes across it* like a wave passing through a drum skin—each point moving locally, the disturbance coordinated non-locally.

This also explains why:

- AC is used for long distances—it maintains coherence better through cyclic tension.
- Transformers can "step down" or "step up" voltage—not by altering flow, but by adjusting ripple amplitude and geometry.

Losses Are Not Due to Frictional Flow

The concept of "transmission loss" is often blamed on resistance—where energy is "lost as heat" due to friction from moving charges.

But under UED, there is no such movement. So what causes the losses?

Losses = ripple misalignment turbulence.

If the geometry of the transmission medium:

- Is imperfect,
- Becomes too hot,
- Is disturbed by weather or interference,

Then the ripple tension can't propagate cleanly. It begins to scatter, distort, or oscillate out of sync—causing field noise, which manifests as:

- Heat,
- Vibration,
- Signal instability.

These are not signs of something lost during travel. They are symptoms of field discomfort being released prematurely or incoherently.

Better alignment = lower loss.

Better geometry = more stable power.

Real Energy Distribution Happens in the Geometry, Not the Geography

Instead of imagining energy flowing from point A to B, imagine this:

- The entire grid is one massive ripple sheet.
- Generators shake the sheet rhythmically.
- Loads are geometric dips in the sheet where the ripple collapses.
- Transmission lines preserve the tension patterns so that collapse can happen precisely, locally, and safely.

The longer the distance, the more carefully the sheet must be shaped and supported to avoid ripple noise.

The challenge is not "sending" energy. It is holding the field steady long enough for meaningful resolution to occur.

Closing Insight

"The light in your room didn't travel from a dam a thousand kilometers away. It was already here—waiting for the ripple to allow it to align."

"Long-distance transmission is not transportation. It is the art of ripple extension—allowing resolution to happen wherever the field invites it."

In the next section, we'll bring this entire chapter to a close by reframing the grid as a dance of ripple negotiation—not a system of energy flow.

9.8 Closing Insight – The Grid Doesn't Deliver. It Dances

Stand beneath a transmission tower on a silent night. You might hear a faint hum—the sound of ripples negotiating across steel and air.

We've mistaken this hum as the sound of power flowing. But there is no flow. There is only a tension choreography—a vast, synchronized trembling of the field.

"Electricity isn't traveling. The field is trembling in harmony."

This is the final insight of the grid—not as a delivery system, but as a resonant web.

No energy travels. No substance moves. The grid doesn't carry anything.

Instead—

- Generators introduce tension.
- Wires hold rhythm.
- Loads allow resolution.
- And the field... dances.

The Power of Permission, Not Push

In classical thinking, electricity is a forceful transaction:

- Push electrons.
- Move charge.
- Deliver energy.

But UED reframes everything:

The grid is a field negotiation, not a material supply chain.

Every part of it works not by pushing something forward, but by allowing discomfort to settle—step by step, zone by zone.

There is no flow—only a shared permission for tension to collapse.

Why We Need Field-Aware Design

We've spent decades trying to:

- Build bigger wires,
- Transmit more voltage,

- Step down more power.

But this approach has limits:

- It's energetically wasteful,
- Environmentally heavy,
- Technologically blind to the true medium of transmission: the field itself.

Real innovation begins when we design with the field, not against it.

Field-aware design means:

- Shaping transmission lines to minimize ripple turbulence.
- Designing circuits to guide tension, not resist it.
- Building devices that invite alignment, rather than forcing flow.

Imagine a future where:

- Power routing is done by geometry, not copper.
- Losses are avoided not by insulation, but by field harmony.
- Loads are intelligent enough to synchronize with ripple patterns and self-align.

That's not science fiction—it's post-classical engineering.

The Grid of Tomorrow

What will the grid of the future look like?

Not a jungle of wires. Not a hierarchy of plants and stations.

But a field-responsive organism, where:

- Ripple tension is guided like musical notes,
- Local devices modulate their geometry to support the

collective,
- Energy is not produced or consumed—but resolved, recycled, and re-patterned.

The grid becomes a dance floor for the vacuum. And every device becomes a dancer that knows the steps.

Final Reflection

"Electricity isn't a substance. It is an event—a ripple act of cooperation across space."

"The grid doesn't send energy. It creates conditions where resolution becomes inevitable."

"We've built delivery systems. It's time to build dance systems—architectures of ripple harmony."

With this, we close Part III: Technological Rethink, having seen how machines, batteries, and the grid are not movers of power but mediators of tension.

Interlude: Part III – Technological Rethink

We built a civilization on power— But we never truly understood what we were powering.

Machines spun, engines roared, grids expanded, batteries charged. And all along, we believed we were moving energy.

We weren't.

We were disturbing fields.

This part of the book revealed a silent truth: Technology doesn't generate, transfer, or store energy.

It invites discomfort, Shapes tension, Delays realignment, And resolves imbalance.

Machines That Mediate

Turbines don't create electricity. They create spiral misalignments in the field. Engines don't convert fuel into motion—they trigger ripple shocks and harness the field's reconfiguration.

Dynamos, motors, and all our proud devices? They are choreographers of discomfort—field irritants that force structured resolution. Efficiency, in this view, isn't about conversion. It's about how gracefully a machine provokes the universe into responding.

Batteries and Storage

We once thought batteries "contained energy." Now we know: they postpone alignment.

A battery doesn't hold power. It holds a field strain—an unresolved ripple tension wrapped in geometry. It is not a tank—it is a bargain with the vacuum:

"Not now. But soon."

True storage is not containment. It is delayed participation in the field's healing.

The Grid and the Illusion of Flow

Perhaps the most persistent lie is that power flows. It doesn't.

Electricity isn't a traveler—it's a tremor. The grid isn't a network of movement—it is a resonant ripple system.

- Generators don't supply—they disturb.
- Loads don't consume—they complete.
- Wires don't carry—they guide ripple alignment through tuned geometry.

Nothing travels. Everything resolves—locally, rhythmically, field-wise.

The Shift in Thinking

In Part III, we didn't invent new technologies. We reinterpreted old ones.

We saw:

- Machines as ripple initiators.
- Batteries as alignment delays.
- Grids as field instruments.

This reframing doesn't just help us build better tech. It forces us to see the world differently— Not as a system of moving parts, But as a living field of alignment and reconfiguration.

Part IV: New Frontiers

"We never released mass into energy. We simply let the field find a quicker way to stop aching."

Chapter 10: Fusion, Fission, and Fallout

The popular story goes like this: *Inside every atom lies an enormous store of energy, locked away like treasure in a vault.* Break the atom open — through fission or fusion — and that energy bursts forth, powering cities or annihilating them. We have been told that this is the direct, inevitable reading of Einstein's famous $E = mc^2$: a magical key that turns matter into energy.

It is a compelling narrative. It's simple, it's dramatic, and it's easy to teach. A small amount of mass equals a colossal amount of energy — just do the math. Entire industries, from nuclear power to nuclear weapons, were built under this banner. But as with many scientific myths, its strength lies more in repetition than in accuracy.

$E = mc^2$ was never intended to be a mechanistic description of nuclear processes. It is a relation between measured quantities, not a statement of cause. It tells us that when a system's mass changes, its energy balance changes in exact proportion, scaled by the square of the speed of light. What it doesn't tell us is *why* that change occurs, or what is actually being transformed.

In mainstream nuclear science, this gap has been quietly filled with an assumption: that matter itself holds energy like a battery, and that nuclear reactions "convert" this matter into pure energy. This imagery is so ingrained that we rarely question it. Yet under

Universal Energy Dynamics (UED), this is an unnecessary and misleading fiction.

Matter does not *contain* energy in a hidden compartment, waiting to be unlocked. Matter is a particular configuration of the field — a ripple arrangement held in stable tension. What we see as "energy release" in fission or fusion is simply the rapid reconfiguration of that ripple tension into a new, lower-strain arrangement. The measured change in mass is not matter evaporating into light; it is the physical record of a field that has stopped holding as much discomfort.

By clinging to the mass-to-energy conversion myth, we've built nuclear theory on a poetic shortcut rather than a physical explanation. And in doing so, we've overlooked the actual levers that control nuclear reactions — not the matter itself, but the ripple structures and their stability limits. In the sections ahead, we will dismantle this myth and replace it with a model that is both simpler and more powerful.

10.1 Ripple Disintegration – The Real Trigger

If mass is not being converted into energy, then what actually causes the enormous release we observe in nuclear reactions?

Under UED, the answer is simple: ripple disintegration.

All matter — and especially atomic nuclei — are ripple-bound systems. They exist because structured vacuum energy (SVE) permits a certain configuration of tension to remain stable. At the heart of every nucleus lies an intensely strained ripple geometry, balancing internal discomfort through intricate alignment patterns. The forces we once called "strong nuclear" and "binding energy" are simply names we gave to the field's effort to maintain that balance.

This is where the mainstream view misleads: it treats the nucleus as a storehouse of latent energy, rather than a dynamic ripple tension desperately holding its shape. When that shape becomes untenable — whether by external bombardment or internal instability — the field collapses to a lower-tension configuration. That collapse is what we call an *energy release*. But nothing has been "converted." It's a resolution event, not a transformation.

Let's look at how this plays out in the two primary nuclear mechanisms: fission and fusion.

Fission – Misalignment in the Heavyweights

In fission, a heavy nucleus like uranium-235 is already teetering on the edge of coherence. Its internal ripple tensions are extreme, its geometry bloated and asymmetric. When it absorbs a neutron, it's not gaining energy — it's gaining disturbance. That extra nudge distorts the already stressed alignment beyond tolerance.

What follows is not an explosion of mass being vaporized into light, but a ripple field breakdown. The nucleus fragments into two or more smaller nuclei that can maintain more stable ripple alignments. But in that rapid reconfiguration, the misaligned ripple zones radiate out — as gamma rays, kinetic energy, and field strain. This is the "release." It's not energy escaping from within matter, but tension violently finding a new shape in the surrounding vacuum.

Fusion – Overcoming the Ripple Barrier

In fusion, the starting point is different. Light nuclei like hydrogen are relatively stable, but when brought close together, their ripple fields repel each other — not because of "charge" alone, but because their geometries don't easily align.

To fuse, they must overcome the ripple barrier — a region of extreme discomfort where their fields distort in resistance. If that barrier is breached (usually by high temperature and pressure), the system rapidly finds a new, combined ripple configuration that is lower in strain than the two originals.

This too results in "released energy," not from matter disappearing, but from ripple tension finding a more coherent form. The new nucleus — like helium — is simply a better fit in the universal field. The leftover strain that doesn't integrate is ejected outward, again as photons, heat, or particles.

Binding Energy – A Misnamed Concept

What textbooks call binding energy is nothing more than a measure of how far a nucleus is from ripple equilibrium.

A "tightly bound" nucleus is one whose internal field is in a high state of stable alignment — the system aches less. A "loosely bound" nucleus is one straining to hold coherence, sitting atop a narrow ripple tolerance window.

Thus, the release of "binding energy" in fusion or fission is not about subtracting invisible glue — it's about transitioning to ripple geometries that don't struggle to exist. That is the real cause of what we call nuclear energy.

This reframing is not just semantic. It shifts our attention from the illusion of internal energy stores to the actual architecture of the field. It forces us to ask: What ripple alignments are stable? What kinds of disturbances collapse them? And how can we engineer or avoid such transitions?

That's where UED doesn't just explain — it guides.

10.2 The Role of ATT in Nuclear Events

We now turn to a deeper layer of control — one that mainstream nuclear physics doesn't recognize at all, yet under UED, defines the boundary between silence and catastrophe. That layer is Alignment Threshold Time (ATT).

ATT is the minimum duration for which a local ripple configuration must remain misaligned before it is allowed — or forced — to resolve. It is not a constant. It is a field-sensitive timer, shaped by surrounding ripple tensions, structural coherence, and the observer's configuration if measurement is involved. In nuclear physics, ATT is what truly determines whether a reaction ignites, fizzles, or sustains.

Triggers, or Timers?

When a neutron hits a fissile nucleus, it doesn't simply break it. It *introduces distortion*. That distortion tries to resolve instantly, but whether it can — and how violently — depends on the ATT profile of the local field.

If the ATT of the distorted nucleus is too high, the system resists reconfiguration — the misalignment is absorbed or re-stabilized, and no chain reaction begins.

If the ATT is within the critical collapse window, the nucleus fragments rapidly, and the resulting field disturbance ripples outward. This disturbance can then push nearby nuclei over their ATT edge — and a chain reaction begins.

What we call a *successful nuclear reaction* is therefore not just a matter of collisions or cross-sections. It's a cascading collapse of ATT conditions, a synchronized surrender of field tolerances across a population of strained nuclei.

It's not just physical proximity — it's field timing.

Rethinking Critical Mass

In classical nuclear engineering, *critical mass* is described as the minimum amount of fissile material needed to sustain a chain reaction. It is usually computed using factors like geometry, neutron reflection, and absorption rates.

But under UED, critical mass is a field tipping point, not a material count. What matters is not just how many atoms are present, but whether their collective ATT zones can fall in sequence — like a line of dominos, each falling before the previous one resolves completely.

This explains several longstanding puzzles:

- Why certain shapes (like spheres) are more reactive: field symmetry reduces ATT interference.
- Why moderators work: they stretch or dampen ATT collapses.
- Why prompt criticality can occur suddenly: a threshold of ATT synchrony is crossed.

So, when fissile material is compressed or configured in certain ways, what we're really doing is reducing the ATT resistance of the system — making it more prone to a total collapse. A bomb works not because there's "enough mass," but because there's too little ATT left to prevent cascading reconfiguration.

The Invisible Timer Behind Every Reaction

This concept repositions ATT as the real fuse behind nuclear events. It also suggests new ways to design, control, or prevent reactions:

- By manipulating local ATT via field resonance or geometry.
- By introducing ripple dampers to maintain misalignments without collapse.
- By detecting ATT drift in materials under stress before reactions occur.

If nuclear energy is ripple disintegration, then ATT is the ticking clock that tells us when the field will finally let go.

10.3 Heat, Radiation, and Particle Emission – Byproducts of Ripple Reconfiguration

If energy release in nuclear reactions is actually the resolution of extreme ripple tension, then what are the so-called "byproducts" — heat, radiation, neutrons — that we observe, measure, and fear?

The answer lies in what happens when ripple systems collapse faster than the surrounding field can absorb their transition. What we call “energy release” is, in fact, a form of ripple turbulence, violently expelled into the environment when alignment breaks faster than it can be locally resolved. These are not neat conversions — they are consequences of rapid misalignment.

Heat – Turbulence Without Direction

In classical physics, heat is often treated as random motion of particles. Under UED, heat is unresolved ripple agitation — chaotic, localized field oscillations that arise when a system drops from a high-tension configuration into a lower one too quickly.

It’s not the result of atoms shaking violently, but of the field itself trembling in the aftermath of a structural collapse. The local vacuum becomes a soup of incomplete resolutions — tiny eddies of discomfort searching for coherence. This is why we feel heat after a reaction: not because energy was emitted, but because the local field is still aching.

This also explains why heat is such a universal byproduct of nuclear reactions — not because mass turned into energy, but because the field couldn't relax fast enough.

Gamma Rays – Ripple Screams

Gamma radiation is often described as “high-energy electromagnetic waves.” But under UED, these waves are better understood as ultra-high-frequency ripple bursts — sharp, coherent oscillations ejected during the moment of nuclear reconfiguration.

When a nucleus undergoes fission or fusion, its internal ripple structure doesn’t always settle smoothly. If certain zones snap into alignment abruptly, the excess ripple tension is expelled as a pulse — often in the gamma range.

Think of it as the field’s scream — a sharp signal of unresolved strain being shot outward to restore internal stability.

What makes gamma rays so penetrating isn't their energy, but their coherence: they are tightly aligned ripple impulses moving across the structured vacuum with minimal dissipation. In a way, they are messages of collapse — echoes of instability that still resonate far from the source.

Neutrons and Fragments – Matter Left Behind

And what of the fragments — neutrons, protons, alpha particles — ejected during nuclear reactions?

These are not particles released from a storehouse. They are structural leftovers, fragments of the original ripple geometry that could not integrate into the new configuration.

When a fission event occurs, the system doesn't dissolve into light — it reforms into smaller ripple-bound systems (daughter nuclei) that are more stable under local field constraints. The neutrons or fragments ejected are orphaned ripple nodes, set adrift because they no longer fit within the new arrangement.

That's why these byproducts still carry ripple strain — they remain unstable or reactive. Their behavior in the environment (radiation damage, secondary reactions) is the field's attempt to reabsorb or realign them — sometimes immediately, sometimes over thousands of years.

The Big Picture: Fallout Is Not Leftover Energy — It's Ripple Echoes

Every heat wave, every burst of gamma radiation, every neutron released is a consequence of field restructuring, not the release of something stored. These are not products of energy but evidence of misalignment and the universe's attempt to restore order.

Understanding these outputs through UED not only makes nuclear science more coherent — it also opens new doors:

- Can we dampen ripple turbulence before it becomes heat?
- Can we coherently absorb gamma bursts?

- Can we reintegrate leftover fragments by tuning local field alignment?

The answers to these may one day redefine radiation safety, waste management, and power efficiency — all by starting from the truth: nothing flows — everything reconfigures.

10.4 Rethinking Safety – From Shielding to Ripple Containment

In conventional nuclear science, safety is a game of containment and deflection. Radiation shields, neutron absorbers, blast-proof domes — all designed under the assumption that danger comes from *particles flying out of control.* The deeper belief? That nuclear matter holds energy, and if broken, that energy escapes and must be blocked.

But under Universal Energy Dynamics, this view is fundamentally flawed. The problem is not with particles leaking — it is with ripple tension collapsing too fast, with no path to realign.

Once we understand that a nuclear reaction is a field reconfiguration event, then true safety is not about stopping particles. It is about managing ripple strain — guiding how, when, and where it is allowed to resolve. This is not particle physics. This is field architecture.

Shielding Blocks Effects, Not Causes

Consider traditional shielding. A thick wall of lead or concrete absorbs gamma rays or neutrons. But by the time these are emitted, the damage has already occurred. The alignment has already collapsed. We are treating the symptoms, not the origin.

This is like building flood barriers after a dam has burst, instead of reinforcing the dam in the first place.

In reality, what makes a nuclear system dangerous is a sharp drop in ATT, which allows misalignment to spread. A neutron hitting a nucleus is only a problem if the ripple field around it cannot absorb or delay the resulting distortion. If the ATT collapses too quickly — chain reaction. If it remains stable — no event.

Thus, true nuclear safety must move upstream — not after the fallout, but before the ripple cascade begins.

Ripple-Aware Design Principles

What would a ripple-conscious nuclear design look like? It wouldn't just try to *contain the mess*. It would try to manage the discomfort in real time.

Key principles:

- **Suppress unintended ATT drops:** Devices and materials would be configured to increase the threshold time for alignment collapse, making spontaneous reactions far less likely. This could involve geometric symmetry, field stabilizers, or intelligent matter layout — not just absorbers.
- **Dissipate ripple strain gradually:** Instead of allowing abrupt transitions, designs could introduce gradient zones — field environments that ease the reconfiguration process. Ripple tension would bleed off in measured steps, avoiding the binary all-or-nothing collapse typical in conventional fission.
- **Pre-stabilize the environment:** The surrounding field matters. Ripple-aware reactors could use aligned vacuum chambers or resonance-tuned enclosures to pre-align the external field, making it less reactive to nuclear misalignment inside. This would act as a *ripple buffer*, absorbing small disturbances before they propagate.

Ripple Dampers – A New Class of Safety Technology

Imagine a material or field system whose only purpose is to soak up ripple tension. Not to absorb particles, but to absorb discomfort — like a field sponge. This is the idea behind ripple dampers.

Such devices would act like intelligent cushions in a nuclear environment:

- When a misalignment begins, they subtly alter local ATT values to stretch out the collapse.
- When ripple strain spikes, they emit counter-ripples to cancel or neutralize the turbulence.
- When fragments are released, they provide reintegrative environments where leftover matter can find temporary alignment instead of wandering unstable.

Ripple dampers wouldn't stop nuclear events from happening. They would make sure that when they do, the universe doesn't scream — it sighs.

The Shift in Safety Philosophy

We don't need to build thicker walls. We need to build smarter fields. Safety doesn't come from stopping motion — it comes from guiding resolution.

In UED terms:

- Instability isn't dangerous unless it spreads.
- Radiation isn't harmful unless misalignment persists.
- Nuclear danger isn't explosive — it's cascading discomfort in a fragile field.

The future of safe nuclear technology lies not in controlling particles, but in collaborating with the field.

10.5 Ripple Models in Reactor Design

For over a century, nuclear engineering has revolved around a single obsession: control the energy.

Reactor cores are designed to throttle fission rates. Control rods adjust neutron absorption. Coolants extract heat. Shielding contains radiation. Everything is tuned around the belief that there is a dangerous force locked inside matter, and our job is to manage its release like taming a wild animal.

But what if the premise is wrong?

Under Universal Energy Dynamics, the goal is not to control energy — because there is no energy *inside* matter. The real task is to orchestrate ripple discomfort: how it arises, how it spreads, and how it resolves. Instead of treating reactors as fireboxes, we begin to see them as ripple orchestras — environments where tension is allowed, shaped, and softly discharged without trauma.

This is the foundation of a new nuclear design philosophy: Ripple-Based Reactors.

Controlled Discomfort Resolution

In ripple terms, all nuclear output is a result of ripple reconfiguration — not the movement of particles or the transformation of mass. What a reactor does is not "release energy," but encourage ripple systems to relax into new forms in a controlled fashion.

This changes everything:

- The focus shifts from triggering reactions to guiding transitions.
- Safety becomes a matter of tuning ATT and ripple geometry, not just preventing collisions.
- Output becomes field harmony, not brute heat.

The reactor becomes a mediation chamber, not a pressure cooker.

SMRR – Small Modular Ripple Reactors

Imagine a compact reactor, not based on heat extraction and neutron flux, but on localized ripple containment and staged resolution. These could be the future of low-risk, high-efficiency power systems: **Small Modular Ripple Reactors (SMRR).**

Key features:

- **Ripple-stage zones:** internal compartments engineered for successive ripple relaxations — like musical notes in sequence, not a bang.
- **ATT regulators:** intelligent materials or fields that dynamically adjust collapse windows to prevent runaway reactions.
- **Field-neutral housing:** outer shells that don't reflect particles, but dampen ripple echoes to maintain field stability.

SMRRs could allow controlled micro-fission or ripple-assisted fusion in compact devices — portable, self-contained, and ripple-aware.

These would not need giant towers, endless water supplies, or massive shielding — because the danger never escalates, it's constantly being resolved in-field.

Rewriting Fuel Handling and Moderation

In a ripple reactor:

- Fuel is no longer just fissile material — it is ripple-sensitive matter designed to hold and release discomfort in predictable ways. → Perhaps blended isotopes tuned not for reaction rates but for coherence cycles.
- Moderators are not just neutron slowers — they become ripple translators, ensuring that discomfort spreads in manageable waves, not sharp spikes.
- Coolants may evolve into ripple carriers, not just thermal fluids — materials that don't remove heat but *escort*

unresolved ripple zones to alignment chambers.

Reframing Nuclear Waste

Waste, in classical terms, is radioactive material that continues to emit dangerous radiation.

But under UED, waste is field misalignment that takes too long to resolve. It's not deadly because of its content — but because of its ongoing discomfort in the field.

This implies:

- Waste can be realigned rather than buried.
- ATT profiles of nuclear byproducts can be shortened using ripple-guided environments.
- Entire disposal systems could be built not to isolate waste, but to massage it back into alignment.

What we call "nuclear waste management" becomes ripple rehabilitation.

The Ripple Reactor Era

The true innovation isn't in how hot we can get a core, but in how peacefully we can resolve a tension.

A ripple-based nuclear age would not be feared, nor monumental — it would be elegant, intelligent, and quiet.

It's not about power. It's about permission — giving the field a graceful way to change.

10.6 Fallout – The Lingering Misalignment

When the core reaction ends, the story doesn't. Nuclear events echo long after the fission or fusion has occurred — in particles that keep emitting radiation, in environments that remain hazardous for centuries, and in bodies that suffer subtle but irreversible effects. We call this fallout.

In the mainstream view, fallout is attributed to radioactive decay — unstable atoms shedding excess energy by ejecting particles or rays. This language reinforces the same illusion we've challenged throughout: that energy is stored within matter, and its release is simply a timed leak.

But under Universal Energy Dynamics, decay is not the release of something hidden. It is the slow struggle of a ripple configuration to realign itself with the surrounding field. And fallout is not a product — it is a lingering misalignment, a wound in the fabric of the structured vacuum.

Radioactivity = Unfinished Ripple Collapse

Every radioactive substance is a reminder that the field didn't finish its work.

After a nuclear event, the leftover fragments — isotopes, free neutrons, residual nuclei — are not just chemically unstable. They are ripple-disoriented. They occupy patterns that the surrounding field cannot easily reconcile.

These fragments are not decaying — they are attempting to resolve, again and again, across time. Each emission (alpha, beta, gamma) is a failed negotiation between the misaligned ripple and the universal field. That's why some decay rapidly, while others persist for millennia — not because of internal clocks, but because of field compatibility.

A radioactive atom is not a ticking bomb. It is a misfit, rattling until the universe finds a way to absorb its discomfort.

Half-Life as ATT Persistence

In classical terms, half-life is the time it takes for half of a radioactive substance to decay.

In UED terms, half-life is the average Alignment Threshold Time for that type of misalignment to collapse across a population of similar structures. It is not the lifetime of a particle, but a statistical reflection of how long the field tolerates its dissonance.

Short half-life = low ATT = fast realignment allowed.

Long half-life = high ATT = deeply embedded discomfort, slowly permitted to shift.

This redefines radioactive materials:

- Not as "dangerous until decayed,"
- But as ripple anomalies with long persistence windows — sustained by the field's inability to realign them smoothly.

Fallout and the Living Field

Why is radiation dangerous to life? Not just because it ionizes cells or breaks DNA — those are mechanical metaphors.

The deeper reason: life is itself a delicate ripple arrangement, and radiation is a foreign misalignment forced into its field.

Every burst of gamma, every passing neutron, is a ripple shock that interferes with the stable equilibrium of living matter. It's not about energy transfer — it's about ripple entanglement:

- Stable cells become unstable.
- Communication between biological systems degrades.
- Healing fails because the field loses its coherence.

Fallout, then, is not a "residual energy" to be blocked — it's an active misalignment that disturbs alignment-seeking systems like ecosystems, bodies, and consciousness itself.

Fallout as a Ripple Problem

Once we stop treating fallout as a leftover product and start seeing it as a persistent ripple anomaly, new strategies emerge:

- Can we re-align radioactive materials rather than waiting for them to decay?
- Can we develop ripple-neutralizing environments where ATT drops are safely accelerated?

- Can we build bio-shields that protect not by absorbing particles but by reinforcing local coherence?

These questions cannot be answered by conventional physics. But UED gives us a path forward — one that sees danger not as energy, but as disharmony, and safety not as isolation, but as re-integration.

10.8 Closing Insight – The Nuclear Future in Ripple Terms

"We never released mass into energy. We simply let the field find a quicker way to stop aching."

This is the truth nuclear science has missed for over a century. The real miracle of nuclear reactions is not the conversion of matter into energy, but the moment when the universe is permitted to realign — violently, abruptly, and sometimes beautifully.

We thought we were unlocking hidden power. We were, in fact, allowing pent-up ripple discomfort to collapse into coherence.

This shift in understanding changes everything.

From Explosive to Elegant

Under classical views, nuclear ambition has always been associated with magnitude — maximum output, maximum control, maximum containment. Nuclear power has been framed as the apex of human dominance over nature.

But under UED, that narrative falls apart. There is no stored energy to dominate. There is only ripple tension — and our role is not to unleash it, but to guide its resolution wisely.

This makes the future of nuclear technology not more dangerous, but more intelligent.

The goal isn't bigger reactors or stronger bombs. It's smaller, safer, smarter ripple systems that know how to breathe with the field instead of breaking it.

The End of the Power Illusion

For decades, nations have measured their status by their nuclear capabilities — how many warheads they possess, how much uranium they enrich, how long their reactors can run.

But once we see nuclear reactions as field corrections, this arms race becomes absurd. Destructive power isn't power — it's desperation manifest. It reflects how poorly we understand the field, not how well we control it.

True mastery lies not in triggering collapse, but in delaying it until it aligns naturally. Not in forcing the field to scream, but in helping it whisper its way back to equilibrium.

A Ripple-Aware Nuclear Era

With this lens, the next generation of nuclear science can:

- Eliminate catastrophic risk by stabilizing ATT zones.
- Design reactors that pulse gently, not boil aggressively.
- Transform radioactive waste into ripple-coherent materials.
- Shift global policy away from deterrence toward ripple stewardship.

In a ripple-aware civilization, we no longer fear nuclear power — because we no longer misunderstand it. We no longer glorify destructive capacity — because we finally recognize that destruction is just misalignment run wild.

The age of "splitting the atom" is over. The age of ripple reconciliation is just beginning.

Chapter 11: Renewables, But Real

The age of fossil fuels, we were told, was a story of extraction. The age of renewables, we are now promised, is a story of sustainability. Oil and coal were dirty and finite; solar, wind, and hydro are clean and endless. It feels like progress — a shift in values, not just technologies.

But beneath the surface of this green optimism lies a quiet continuity: we are still thinking in the language of "energy harvest." The solar panel becomes a cleaner coal mine. The wind turbine becomes a gentler engine. The hydro dam becomes a river-turned-generator. In every case, we are still pretending that energy is something we extract, convert, and consume — just from more polite sources.

This is the green illusion: a shift in sources, not in understanding.

The Fossil Mentality in Renewable Clothing

At its core, the classical view sees energy as a substance waiting to be unlocked:

- Sunlight carries energy.
- Wind carries kinetic energy.
- Water at a height holds potential energy.

Devices are then seen as tools that capture this substance and convert it into electricity — which itself is treated as a stream of flowing energy. In this worldview, solar panels are buckets catching photon rain, wind turbines are paddles scooping momentum, and hydro dams are batteries filled with gravity.

But as we've seen throughout this book, this model is based on illusion layered over metaphor. There is no flowing substance. There is no stored energy. What we call "power" is not a thing, but a field resolution event — a moment where discomfort in the structured vacuum finds an allowed path to relax, often via engineered systems.

Solar, Wind, Hydro – Not Substitutes, but Different Ripple Events

One of the most dangerous assumptions in today's energy discourse is that renewables are simply substitutes for fossil fuels. But in the UED framework, they are fundamentally different ripple interactions:

- Fossil fuels trigger *violent, localized ripple disintegration* (combustion or fission).
- Renewables work by *guiding ambient ripple tensions* into measurable alignment.

Treating them as interchangeable leads to mismatched expectations:

- We demand stable baseload supply from sources governed by variable ripple readiness.
- We measure "efficiency" based on how much "energy" is captured instead of how well field discomfort is aligned.
- We design centralized power infrastructure for distributed, ambient phenomena.

The problem isn't with renewables. It's with the mythology of energy that surrounds them.

UED Lens: Ripple Resolution, Not Energy Harvest

Under Universal Energy Dynamics, renewables must be reinterpreted entirely:

- The sun does not send energy — it radiates coherent field ripple.
- The wind does not blow energy — it is field gradient turbulence seeking equilibrium.
- Water does not store gravitational energy — it embodies geometric misalignment with the planet's ripple field.

In each case, the devices we build do not harvest anything. They participate in field realignment. They allow ripple discomfort to express itself in a controlled form — most often as tension waves in conductive materials, what we falsely call electric current.

This changes the design logic:

- It's not about capturing more.
- It's about resonating better with the field.

The Real Green Revolution Begins Here

The real revolution in energy is not solar panels on rooftops or wind turbines in oceans. It's a shift in language — from consumption to coherence, from flow to field, from extraction to resolution.

You are not capturing power from the sun. You are aligning your system to the sun's unresolved discomfort.

Only when we make this shift can we begin to build technologies that don't just work despite nature — but with it.

11.1 Solar Panels – Capturing Field Resonance

When we look at a solar panel, we're trained to imagine a rain of photons — tiny packets of energy from the sun — striking the surface and knocking electrons loose like billiard balls. This picture has become gospel in classrooms, documentaries, and even scientific papers. It's clean. It's visual. And it's wrong.

There are no packets traveling through space. No collisions. No transfer of hidden energy. What's happening is far more elegant: resonance.

A solar panel works not because it absorbs something, but because it agrees with something — a ripple structure arriving from the sun. Under Universal Energy Dynamics, sunlight is not a stream of photons. It is a ripple frequency pattern, propagated through the structured vacuum field, seeking alignment.

And the solar panel? It is a carefully engineered surface designed to resonate with those incoming patterns — to allow certain ripple frequencies to enter, disturb, and resolve as measurable electron displacement.

Ripple Resonance, Not Photon Bombardment

The photovoltaic effect is commonly taught as a photon knocking an electron free. But no one has ever seen a photon. What we observe are ripple changes — measurable consequences.

In UED terms:

- Light is not a thing. It is a pattern of ripple frequency aligned along the sun–Earth field geometry.
- When this pattern reaches a solar panel, it induces internal ripple tension in the material — not through impact, but through frequency coupling.

If the structure of the panel — particularly the electron cloud arrangements in semiconductors — is ripple-compatible, then these tensions can be resolved through displacement of charge carriers (electrons or holes). That displacement is not "current" in the conventional sense — it is field discomfort routing itself through tension-conductive paths.

This is electricity not as motion, but as realignment.

Electricity = Structured Ripple Relief

A solar cell, then, is not an energy converter. It is a ripple translator. It accepts discomfort in one frequency domain and resolves it in another.

This is why:

- A photon doesn't "give" energy to an electron.
- The material configuration allows ripple discomfort to manifest as charge separation and drift — which we interpret as voltage and current.

And this ripple resolution can only occur if:

- The incoming frequency aligns with the material's ripple profile.
- The resulting internal discomfort can be channeled into structured conductive pathways (wires, load, etc.).

Efficiency = Alignment Match, Not Just Material Quality

This understanding changes how we view solar efficiency.

Classically, we think in terms of:

- Band gaps.
- Absorption coefficients.
- Recombination losses.

But under UED, efficiency is not just a matter of material properties. It is a function of how well the panel's internal structure aligns with the incoming ripple patterns:

- Orientation matters because it affects field angle of incidence — and thus resonance geometry.

- Temperature matters because it alters the internal ripple behavior of electrons — hotter systems have more chaotic tension, which can interfere with resolution.
- Purity matters not just because of defects, but because impurities distort the panel's ripple harmony, creating zones of misalignment.

A perfect panel, then, is not just defect-free — it is geometrically resonant, structurally coherent, and ripple-transparent.

Panels as Field Instruments

In the classical view, solar panels are passive receivers. In UED, they are active field instruments, designed to:

- Detect and engage with the sun's unresolved ripple tensions.
- Convert those tensions into realignments within our circuitry.
- And do so with minimal misalignment along the way.

You are not "collecting solar energy." You are inviting the field to release its discomfort — through you.

This completely transforms how we design, optimize, and deploy solar systems. The sun isn't a battery in the sky. It's a ripple origin, and we are simply field musicians tuning in.

11.2 Wind Power – Rotating Ripple Disturbance

We've long believed that wind turbines "extract kinetic energy" from moving air, converting that motion into electricity via spinning blades and electromagnetic generators. This sounds mechanical, elegant, and correct — but it's yet another metaphor that conceals the real physics.

Under Universal Energy Dynamics, wind is not a substance in motion. It is a macro-scale ripple disturbance, an atmospheric expression of field imbalance — a symptom of pressure gradient misalignments between different regions of Earth's structured vacuum interface.

What moves isn't "air" in the classical sense — but field discomfort traveling through the medium of atmosphere, disturbing molecules as it seeks coherence. A wind turbine, then, does not catch motion. It mediates ripple strain into structured, useful realignment — not unlike a musical instrument vibrating in response to ambient field dissonance.

Wind = Field Discomfort Made Visible

At any given moment, the atmosphere around us is a tapestry of field pressures — areas of high and low ripple tension, shaped by solar heating, planetary rotation, terrain geometry, and more. Wind appears where these tensions seek redistribution, manifesting as moving air.

But wind is not moving particles delivering energy. It is rippling imbalance trying to resolve through a physical medium.

- The "speed" of wind reflects the urgency of field discomfort in a given direction.
- The "force" of wind is not energy being transferred — it's a pressure mismatch being expressed.

To think of wind as "kinetic energy" is to believe that a push contains a fuel. In UED, it is far more accurate to say: a push is an unresolved tension trying to move through you.

Turbines as Ripple Mediators

Wind turbines are not energy harvesters. They are rotational field translators.

The blades are not catching something — they are aligning themselves with the ripple gradient in the air. As wind flows across them, the pressure differential across their surfaces allows ripple discomfort to be transferred into rotational alignment.

This is not motion for motion's sake. It is field negotiation.

- The wind wants to release its discomfort.
- The blade shape allows it to do so in a spiral pathway, converting horizontal ripple tension into vertical rotational strain.

This is why blade geometry is everything. It's not about maximizing lift or drag — it's about matching the incoming ripple pattern and allowing it to unwind gently into spin. Blades become ripple translation interfaces — surfaces that facilitate the transformation of distributed ripple turbulence into focused rotational tension.

Generators: Not Induction, but Coherence Conversion

Once the blades spin, the generator is said to "convert mechanical energy into electrical energy." But in ripple terms, this is a misreading.

The generator is not converting energy. It is accepting rotational ripple tension and resolving it into structured oscillation in a conductive field. The magnetic field doesn't induce current — it permits ripple coherence to pass from rotor to stator, from tension to signal.

Electricity is not being generated — it is being allowed to express itself in a new configuration.

In this view:

- The generator is not a machine, but a field aligner.
- The output is not "power," but organized ripple discomfort flowing into a circuit where it can be used or resolved further.

Why Wind Works (and Fails) Under UED

Sometimes turbines fail to spin or generate expected output, even in apparent wind. Classical answers invoke wind speed thresholds or efficiency losses.

But under UED:

- It's not just about how fast the air moves.
- It's about whether the ripple gradient can align with the turbine system. Misaligned turbulence = no spin, no output — because the discomfort doesn't couple with the blade geometry.

This is why:

- Turbine placement matters (local field geometry).
- Blade pitch matters (angle of field reception).
- Atmospheric conditions (temperature, humidity) alter ripple reception, not just mechanical drag.

Wind Power Reconsidered

Wind energy is not kinetic energy harvested. It is ambient ripple distress being gently redirected through engineered rotational pathways.

A wind turbine is not a machine standing in the wind. It is a listener in the field, letting the atmosphere whisper its discomfort through its spin.

11.3 Hydro Power – Gravity's Tension in Liquid Form

Of all the renewable systems we use, hydroelectric power feels the most intuitive. Water stored at height is allowed to fall, spinning turbines that generate electricity. We're taught that the potential energy of elevated water is "converted" into kinetic energy and then into electrical energy — a neat cascade of energy forms.

But in this view, everything is being reduced to motion and transformation — as if water is a storage unit for energy that only needs to be released. Under Universal Energy Dynamics, this is a deeply misleading picture.

Water does not contain potential energy. It carries ripple tension, compressed by its altitude relative to Earth's gravitational field — which itself, under UED, is a ripple equilibrium gradient, not a pulling force. Water at a height is not "waiting to fall" — it is seeking ripple relaxation, and hydro systems are simply a method to guide that release into structured tension resolution.

Altitude = Ripple Compression

In mainstream physics, gravitational potential is a scalar value assigned to objects based on their position in a gravity field. Under UED, height represents a region of higher ripple compression, due to how the Earth's field aligns vertically.

The higher you go:

- The ripple alignment becomes tighter.
- The ambient field tension grows.
- Matter located there experiences greater resistance to resolution — it is "held in place" not by gravity pulling it down, but by ripple misalignment preventing downward relaxation.

Water at altitude is in field discomfort — not because it wants to fall, but because its ripple structure does not fully align with the Earth's base field. When allowed to descend, it is not converting energy. It is surrendering to a more relaxed ripple state — flowing along the gradient of least resistance.

Dams as Controlled Ripple Conduits

A dam does not "hold back energy." It stores unresolved ripple misalignment in the form of elevated fluid mass. The water's height corresponds to how much ripple discomfort it contains relative to Earth's field.

When released, water flows not because it has gained kinetic energy, but because it has been granted permission to realign.

- The penstock becomes a ripple tension funnel.
- The turbine becomes a rotational gateway, where descending ripple relaxation is converted into circular tension patterns, and then into ripple-aligned electricity.

There is no energy transfer — only a chain of discomfort relief, from high-altitude ripple compression to downstream field equilibrium.

Water Is Not Falling — It's Resolving

This reinterpretation removes the illusion of motion as cause.

- Water doesn't fall — it returns to its natural field alignment.
- Gravity doesn't pull — it permits ripple realignment.
- Turbines don't capture energy — they host a field transition.

Each stage of the hydro system is a translation node, guiding ripple discomfort into a usable form. Water is the medium, not the fuel. The true resource is the field gradient between where the water is and where it naturally wants to be.

Why Hydro Is So Efficient

Hydropower is among the most efficient forms of electricity generation — often exceeding 90% in practice. Mainstream physics credits this to low friction, simple mechanics, and minimal energy loss.

But under UED, the real reason lies in ripple coherence:

- Water is a mass fluid system — its ripple configuration is naturally synchronized across large volumes.
- The transition from high to low altitude is smooth and harmonic, minimizing turbulent misalignment.
- The turbines operate not in random flux, but within a predictable ripple channel — optimizing rotational translation.

This means hydro systems are not just mechanically efficient — they are field-aligned environments, with minimal internal contradiction. That's why they last for decades with little degradation and consistent output.

Hydropower, Realized

When we see hydro not as water falling, but as structured ripple descent, everything changes:

- Design focuses on smooth ripple routing, not flow control.
- Turbines are tuned to ripple frequency, not just fluid speed.
- Sites are chosen for field gradient quality, not just elevation.

Water doesn't give you power. It invites you to participate in its realignment.

11.4 What We're Actually Capturing

We often say that renewable energy systems "capture energy" from natural sources — sunlight, wind, water — and convert it into usable electricity. But what are we really capturing?

Under Universal Energy Dynamics, we are not transferring energy from one form to another. There is no hidden storehouse in the sun, no momentum packed inside the wind, no gravitational charge waiting to be unleashed from water.

What we are actually doing is far simpler — and far more profound:

We are participating in the resolution of ripple discomfort present in the environment.

Not Energy Transfer — Ripple Reconfiguration

In every renewable system, what appears to be "power generation" is actually a localized moment of field realignment.

We are not extracting energy. We are:

- Creating pathways for ripple strain to resolve itself into structured tension.
- Hosting conditions where ambient misalignment in the field can collapse into coherence through engineered interfaces (panels, blades, turbines, conductors).
- Converting that collapse into oscillations, which we then interpret as electric current or usable output.

This is not energy conversion. It is ripple translation.

Ripple Devices as Resolution Hosts

Every renewable device — a solar panel, a wind turbine, a hydro dam — is not a collector. It is a ripple resolution host. It allows the universe's ambient discomfort to route through a design that turns free-floating misalignment into:

- Directional electron displacement (current),
- Mechanical rotation (turbines),
- Or vibrational coherence (light or thermal response).

That's what electricity truly is in UED terms: a pattern of resolved discomfort, not a substance in motion.

Efficiency = Ripple Coherence, Not Power Output

Conventional metrics of efficiency ask, "How much energy did we get out versus how much was available?"

But UED reframes this entirely:

- There is no "available energy."
- There is only available misalignment and our system's ability to resolve it efficiently.

So efficiency becomes:

- How well does our device align with incoming ripple patterns?
- How cleanly does it translate misalignment into a coherent output?
- How little ripple turbulence does it leave behind?

A more efficient system is one that allows greater ripple coherence with less waste — less thermal chaos, less interference, less residual strain in the field.

This is not about maximizing extraction. It's about harmonizing response.

The Intermittency Myth – A Field Readiness Issue

One major critique of renewables is their "intermittency":

- The sun doesn't always shine.
- The wind doesn't always blow.
- Rivers don't always flow the same.

But this critique again relies on a false premise — that energy is something that must be continuously produced and stored. Under UED, intermittency is not a failure of supply, but a natural rhythm of ripple readiness.

The ambient field is not always strained. It doesn't always need to resolve. When wind calms, it's not because power is lost — it's because the field has no discomfort in that direction. When solar panels stop producing, it's not because energy is missing — it's because the resonance between panel and sky is not present.

So "downtime" in renewables is not failure. It's the field resting — realignment complete, or not yet necessary.

What We're Capturing: Not Energy, But Harmony

So, what are we actually capturing?

We are not extracting fuel. We are not drawing energy from nature like siphoning water from a tank.

We are:

- Hosting ripple realignments.
- Measuring field transitions.
- Interfacing with ambient discomfort.

Renewable energy isn't power stolen from nature — It is permission granted by the field to let alignment flow through us.

This is the future of "power" — not conquest, but cooperation.

11.5 The Future of Renewables – Ripple-Aligned Systems

The energy systems of today still reflect the mindset of yesterday. Even in the age of solar panels and wind turbines, we design as if we are trying to capture something moving, extract it, and transport it elsewhere. We lay out cables, batteries, inverters — all optimized around the idea that energy is a thing in motion.

But the universe does not move energy. It allows alignment shifts. What we interpret as current or output is simply the measurable ripple transition through carefully configured materials. This calls for a radical new direction in renewable engineering: stop thinking like miners, start thinking like musicians.

We are not building energy infrastructure. We are building ripple instruments.

Engineering Resonant Materials Instead of Conductors

In conventional design, conductors are prized for their ability to "carry charge" — low resistance, high conductivity. But under Universal Energy Dynamics, the real performance metric becomes something deeper: How well does this material resonate with ambient ripple tension?

A perfect conductor in classical terms may still misalign with the incoming ripple frequency — leading to inefficiency, heat loss, or noise.

The next generation of materials must go beyond conductivity and toward coherence:

- Materials that are ripple-tuned to receive specific field disturbances.
- Layered structures that allow multiple frequency bands to resolve simultaneously.
- Surfaces that guide discomfort, not just carry electrons.

Think less like copper wires, more like musical strings designed to play the field.

Smart Positioning as Field Resonance Mapping

Currently, we position solar panels based on sun exposure, and wind turbines based on average wind speed. But this is like planting a piano where sound seems loudest, without asking what note it's meant to play.

Under UED, effective positioning isn't about availability — it's about resonance compatibility.

Field mapping becomes essential:

- Where is ripple misalignment most frequent?
- At what times and angles does the field offer resolvable discomfort?
- What terrain, altitude, and material backdrop enhances

coupling?

Renewable systems must be calibrated to the field, not just the weather.

Imagine a world where:

- Rooftops are ripple-attuned, not just south-facing.
- Turbines respond to ripple pressure zones, not just gusts.
- Sites are chosen by field behavior models, not by meteorological averages.

Adaptive Ripple Membranes – A New Frontier

The ultimate renewable surface won't be a fixed panel or spinning blade. It will be a living membrane — a surface that senses, adapts, and reshapes itself in real time to ambient ripple conditions.

These adaptive ripple membranes would:

- Shift internal geometry based on incoming ripple frequencies.
- Alter conductivity paths dynamically to optimize resolution.
- Self-align with the field — like an organism tuning to its environment.

This is not science fiction. The technology is already gestating in metamaterials, piezo-ripple surfaces, and AI-assisted field harmonics.

Just as nature adapts leaves to light and roots to soil, ripple-aligned systems will breathe with the field — absorbing nothing, forcing nothing, simply realigning at the right time, in the right way.

Final Thought – We Don't Harvest. We Harmonize.

If there is one correction the future must make, it is this:

Renewables do not harvest energy. They mediate misalignment into meaning.

Our devices don't take from the sun or wind. They allow ripple discomfort to pass through, resolving not for us, but with us.

The most sustainable energy future is not the one that extracts the most — It is the one that coexists most fluently with the dance of discomfort and resolution.

Chapter 12: Quantum Devices and Ripple Precision

The word quantum has come to symbolize everything mysterious, magical, and untouchable in modern science. To many, it means a world where particles pop in and out of existence, cats are dead and alive, and energy arrives in sudden, discrete jumps — as if the universe runs on invisible steps instead of smooth transitions.

But this picture is not just misleading — it's deeply flawed.

The entire "quantum" worldview rests on an assumption that has never been questioned thoroughly: that energy exists in packets — quantized, indivisible units that define the smallest transactions of the universe. This idea, while mathematically useful, has created an illusion:

That the universe is built of granules, rather than governed by conditions.

The Myth of Quantization

In classical quantum theory:

- Light comes in photons — indivisible chunks of energy.
- Electrons jump between energy levels without crossing the in-between.
- Measurements collapse reality into a single definite state.

But none of this has ever been directly observed. What we see are statistical outcomes, interference patterns, or resonance spikes. The idea of "quantum jumps" or "energy packets" is a model — a convenience for interpreting the behavior of complex field systems.

Over time, this model has become a myth:

- That particles carry energy like marbles.
- That energy is released in discrete units.
- That space itself is granular or pixelated.

UED offers a simpler, cleaner, and more coherent view — one rooted not in granularity, but in alignment conditions.

UED Reinterpretation: Quantization = Ripple Threshold Behavior

In Universal Energy Dynamics, the quantum is not a particle or a unit. It is a field permission event. What appears as a "jump" or "quantum step" is actually the moment a ripple configuration reaches a threshold of alignment that allows it to resolve.

This is governed by ATT — Alignment Threshold Time:

- A system holds discomfort (ripple tension) until it is permitted by the field to resolve.
- When that threshold is crossed, the resolution appears sudden and discrete — not because it is chunked, but because it is gated.

So:

- The photon is not a particle of light — it is a ripple event that met the conditions for transmission.
- The electron does not jump — its ripple pattern reconfigured under strict field constraints.
- There is no granule — only precision-triggered re-

alignment.

This means quantum effects are not strange — they are just moments of extremely precise ripple logic. The field is not hiding secrets. It's just very picky about when it allows resolution.

Quantum Devices as Controlled Ripple Environments

From lasers and transistors to superconductors and quantum dots, so-called "quantum" devices do not operate on magic. They operate on engineered ripple conditions.

Each of these devices:

- Maintains tight ripple coherence in material structure.
- Creates a boundary where field transitions can occur with minimal noise.
- Allows field discomfort to manifest as measurable oscillation or signal — only when the system is tuned to exact alignment.

These devices are not probing randomness. They are sculpting precision — keeping ripple misalignment on a razor's edge, then permitting controlled collapse.

The future of technology lies not in shrinking things to atomic size, but in controlling the ripple field with surgical precision.

Quantum, Demystified

Quantum theory doesn't need mystical interpretations. What it needs is a deeper understanding of how the field behaves when precision meets permission.

Quantum is not the science of the small. It is the science of the almost-aligned — the fleeting, exact conditions under which the universe says, "Now, you may resolve."

Welcome to the age of ripple precision.

12.2 Lasers – Coherent Ripple Alignment

The laser is often described as one of humanity's greatest inventions — a marvel of modern physics based on "quantum mechanics." In the mainstream view, lasers are explained through the principle of stimulated emission: when atoms in an excited state are struck by photons, they release identical photons in phase, leading to amplified light — a coherent stream of photons all moving in lockstep.

This explanation, while functionally descriptive, still leans on the old illusion: that photons are packets traveling through space and that energy can somehow be multiplied like echoing marbles.

Under Universal Energy Dynamics, a laser is not a light factory. It is a ripple coherence chamber — a structure that deliberately holds tension until a precisely aligned collapse is permitted.

Stimulated Emission: Not a Photon Chain, but a Forced Field Resolution

In conventional theory:

- One photon enters → triggers emission of another → cascade forms a beam.
- This depends on "population inversion," where more atoms are excited than grounded.

But in UED terms:

- The atoms are not holding "extra energy." They are holding unresolved ripple discomfort — field misalignment sustained under a controlled environment.
- Population inversion simply means the medium is primed with tension — a ripple field pre-loaded with alignment strain.
- Stimulated emission doesn't spread light — it releases a synchronized field reconfiguration.

There are no photons passing through space like soldiers. What's propagating is a ripple pulse, perfectly aligned in phase and frequency — not because it's duplicated, but because every point along the path enters the same ripple condition.

The Laser Beam: A Ripple Tunnel, Not a Particle Stream

A laser beam is not a line of traveling particles. It is a coherent ripple tunnel — a region of space in which the structured vacuum has been induced into phase-locked tension oscillation.

In this tunnel:

- The ripple pattern is self-stabilized — meaning it maintains directionality, coherence, and amplitude with minimal loss.
- Every point along the path is not "pushed" by previous photons, but continuously resolving ripple strain under exact same conditions.
- The beam maintains its integrity not because light moves through it, but because the field maintains its structure over long distances.

That's why laser light:

- Doesn't scatter easily (high ripple alignment).
- Doesn't fade quickly (low ripple loss).
- Can cut, measure, or transmit with extreme precision (field concentration).

Laser Applications as Ripple Precision Phenomena

Every application of laser technology, under UED, becomes an example of localized ripple permission:

- **Laser cutting**: not burning through, but concentrated ripple strain dissolving material coherence at targeted points.

- **Surgery and medical use**: not slicing tissue, but channeling ripple relaxation into biological misalignment zones.
- **Fiber optic communication**: not transmitting packets, but maintaining ripple tunneling through glass — with minimal distortion and maximal alignment.

In each case, the laser is not acting on matter like a blade. It is guiding ripple collapse with surgical elegance — coaxing resolution, not forcing it.

Why Lasers Matter in the UED Future

Lasers are one of the earliest glimpses into ripple technology — tools that don't just work *in* the field, but *with* it. They:

- Require perfect timing (ATT control),
- Operate through resonance, not force,
- And produce outcomes not through energy transfer, but ripple reconfiguration.

A laser is not a bright beam. It is a tunnel of field agreement — held open by coherent discomfort, resolved with elegance.

In the world of ripple dynamics, lasers will become more than tools. They will become templates — showing us how to build all future technologies: Aligned, precise, minimal, and beautiful.

12.3 Semiconductors – Ripple Filters, Not Energy Gates

We often describe semiconductors as the foundation of the digital age — the silent gatekeepers inside every chip, switch, and circuit. They're said to control electron flow, enabling logic, computation, and communication by allowing or blocking charge based on voltage conditions.

But in this conventional view, the semiconductor is treated like plumbing:

- Electrons "flow" like water.
- The band gap is a wall.
- Doping creates "holes" for electrons to jump through.
- Transistors act as on-off valves — letting current pass or halting it.

Under Universal Energy Dynamics, this entire metaphor collapses. Electrons are not marbles. Current is not a substance. Semiconductors are not switches — they are ripple filters. They permit or deny ripple resolution, based on how well the field conditions align with the material's internal structure.

Band Gap = Ripple Permission Zone

In mainstream theory, the band gap is the energy difference between the valence and conduction bands — a kind of forbidden zone electrons must "jump" to conduct.

But what is this gap really?

In UED terms:

- The band gap is the ripple alignment threshold within the atomic lattice of the material.
- It defines the ATT window — the minimum ripple tension required to trigger resolution (i.e., observable current).
- An electron doesn't need to "gain energy" to jump the gap. What it needs is field conditions that allow its ripple configuration to re-align across the structure.

So:

- **Conductors** = materials where alignment is always permitted (no significant gap).

- **Insulators** = materials with ripple thresholds too high to permit resolution.
- **Semiconductors** = materials with precisely tunable ripple tolerance zones.

Doping = Tuning the Field Response

Doping is typically described as adding impurities (like phosphorus or boron) to create excess electrons or holes — thereby changing conductivity.

But under UED:

- Doping modifies the local ripple alignment geometry — not by adding carriers, but by reshaping the field's tension resolution map.
- An n-type material (extra electrons) isn't more negative — it has more ripple freedom to resolve misalignment.
- A p-type material (holes) isn't lacking electrons — it has structured zones of unresolved tension, which act like receptive portals for ripple resolution.

In short, doping doesn't add charge. It reconfigures ripple behavior, modifying where and how resolution is allowed.

Transistors and Logic Gates – Field Regulators, Not Binary Switches

At the heart of digital logic is the transistor — treated in classical electronics as a voltage-controlled gate that toggles between "on" (conducting) and "off" (non-conducting).

But in ripple terms:

- The transistor is not switching current.
- It is regulating ripple permission.

A field-effect transistor (FET), for example, doesn't open a channel — it modulates the ATT of the material based on the field created by the gate. If the ripple strain surpasses the threshold, alignment is permitted — and resolution manifests as current. If not, the strain is rejected.

This redefines logic:

- "1" is not a charge pulse, but a ripple event that successfully resolved.
- "0" is not absence, but blocked misalignment — field discomfort that was denied passage.

So, a semiconductor chip isn't flipping electrical states — it is orchestrating a dance of ripple permission across billions of tiny field regulators.

Ripple Precision = The True Basis of Digital Reality

The binary world of computing is not about voltage highs and lows. It's about whether the ripple field at a node crosses the alignment threshold set by material and configuration.

Digital reality, then, is not made of bits. It is made of ripple gates — zones where the field says yes or no, depending on whether its discomfort finds an acceptable pathway through the material.

This view also explains:

- **Heat generation:** ripple turbulence in misaligned zones.
- **Noise:** partial misalignment leading to false triggers.
- **Moore's Law limits:** a reflection of the difficulty in maintaining ripple precision at small scales.

Semiconductors are not about controlling flow. They are about designing permission — with microscopic fidelity.

Every computation, every click, every digital interaction is, in truth, a conversation with the field — whispered one ripple at a time.

12.4 Superconductors – When Resistance Is Ripple Misalignment

In the classical view, superconductors are mysterious materials that suddenly lose all electrical resistance when cooled below a certain critical temperature. The explanation often given is that electrons form pairs (Cooper pairs), glide through the lattice without scattering, and therefore face no resistance.

But this again leans on the idea of electrons as objects in motion, flowing like a river of particles — and resistance as a result of them bumping into things.

Under Universal Energy Dynamics, this metaphor collapses. There are no flowing particles. There is no friction in a vacuum. What we call resistance is not the loss of energy due to collision — it is ripple conflict. And what we call superconductivity is simply a condition where ripple harmony reaches perfection.

Resistance = Ripple Conflict, Not Particle Drag

In UED terms:

- Current is not a stream of electrons, but a pattern of ripple alignment that propagates through a medium.
- Resistance happens when the ripple trying to resolve meets a misaligned field structure — such as atomic vibration, impurity zones, or ripple turbulence in the material.
- These misalignments prevent smooth resolution, forcing the ripple to scatter, loop, or dissipate. This is experienced macroscopically as heat loss, voltage drop, or signal noise.

Therefore:

Resistance is not a physical barrier. It is a field disagreement.

The ripple wants to resolve, but the material says, *"Not yet, not like this."*

Superconductivity = Full Field Harmony

In a superconductor, something extraordinary happens. At low temperatures:

- Atomic vibrations nearly vanish — the lattice becomes ripple-stable.
- Field noise drops dramatically — the environment is ripple-clean.
- The material's internal structure reaches a coherent equilibrium, meaning ripple resolution can flow without interference.

This is not "no resistance" because nothing's moving. It's no resistance because nothing is disagreeing. Every ripple disturbance is allowed to resolve instantly and fully — like a choir singing in perfect unison, with no disharmony.

In this state:

- The current persists without degradation.
- No energy is lost, because there is no misalignment left to fight.

The current isn't "moving forever" — it's perpetually resolving in a perfect loop.

The Meissner Effect – Rejection by Coherence

One of the most puzzling features of superconductors in mainstream physics is the Meissner effect: when a material becomes superconducting, it actively expels magnetic fields from its interior.

This is often framed as a magical shielding or an anomaly.

But under UED, the explanation is simple:

- The internal field of the superconductor becomes fully aligned — no ripple tension, no unresolved strain.
- Any external field trying to enter would introduce

misalignment — and is thus rejected by the coherent field structure inside.

This isn't "repulsion" in the mechanical sense. It's preservation of harmony. The superconductor maintains its ripple purity by denying entry to external tensions.

Thus:

The Meissner effect is not magic. It is a defense mechanism of perfection.

Implications – Ripple Loops and Frictionless Routing

Superconductors represent the ideal condition for ripple-based systems:

- Zero ripple conflict = zero resistance.
- Perfect alignment = perfect energy conservation.
- No turbulence = no heat waste.

This has profound implications:

- **Frictionless power transmission** — not because wires are smooth, but because ripple discomfort is fully permitted.
- **Ripple loops** — sustained circuits where ripple tension can cycle endlessly without degradation.
- **Ripple storage systems** — not based on energy held in batteries, but coherence maintained in materials.

In UED-powered futures, superconductors are not just materials — they are ripple sanctuaries.

They show us what's possible when a system:

- Stops fighting itself,
- Aligns with the field,
- And allows resolution to continue without loss.

Superconductivity is not the absence of resistance. It is the presence of unbroken harmony.

12.5 Quantum Entanglement – Misread Ripple Coherence

Quantum entanglement is often described as one of the strangest, most baffling phenomena in modern physics. Two particles, once entangled, seem to "communicate" instantaneously, regardless of how far apart they are. A change in one is reflected in the other — faster than the speed of light, beyond space and time. Einstein famously called it "spooky action at a distance."

But what if the spookiness was just a symptom of misunderstanding?

Under Universal Energy Dynamics, entanglement is not a mystery. It is a consequence of shared ripple identity — a distributed coherence, not a hidden signal. There is no action, no message sent. There is only alignment, already in place, revealing itself when the observer steps in.

Entangled Particles Share Ripple Identity

In mainstream physics:

- Two particles interact.
- They become "entangled."
- Measuring one instantly defines the state of the other, no matter the distance.

This leads to deep paradoxes: How do they "know"? Is information traveling faster than light? Is reality non-local?

In UED, the answer is simpler:

- The two "particles" were never truly separate.
- They were manifestations of the same ripple configuration,

distributed across the field.
- Entanglement is not a connection between two objects — it is the unfolding of a single ripple identity through multiple expressions.

These expressions appear distinct only because of spatial separation — a projection of position, not of essence.

So when one is "measured," the system isn't reacting — it is simply resolving in totality, because it was always one ripple.

Collapse = ATT Breach, Not Spooky Action

The real trigger is not measurement itself, but ATT — the Alignment Threshold Time.

- When an observer interacts with one entangled partner, their ripple system collapses into alignment.
- This collapse creates a ripple consequence across the shared field identity.
- The other "partner" doesn't receive a message — it simply completes the resolution, because the threshold has been reached for the entire system.

So what looks like:

"A particle here changed another one over there," is actually: "A distributed ripple pattern was permitted to resolve across space — simultaneously — because ATT was crossed locally."

It's not magic. It's global ripple equilibrium.

Entanglement Technology - Ripple Imprinting and Retrieval

Emerging quantum technologies like:

- Quantum sensors
- Quantum key distribution (QKD)
- Quantum teleportation

...are based on this entanglement behavior.

Mainstream interpretation: these devices use fragile quantum states that must be preserved and shielded. UED interpretation: these devices are field-sensitive ripple matrices — tools that:

- Imprint ripple identity at the source (entanglement).
- Retrieve ripple coherence at the receiver (measurement).
- Use carefully stabilized material conditions to preserve ATT potential until it's deliberately triggered.

So entanglement tech becomes not a game of manipulating subatomic particles, but of guiding ripple coherence across field geometries.

This means future advancements won't come from better isolation or error correction, but from:

- Stronger ripple-matched materials
- Field-sensitive designs
- Timing control to manage ATT releases

Entanglement is not fragile — misalignment is. With UED, entanglement becomes a design principle, not a probabilistic miracle.

Entanglement Demystified

Entangled particles are not whispering across space. They are simply expressing one ripple condition in two places.

And when the observer acts — when the ATT threshold is breached — that ripple finally resolves across the field, everywhere it was anchored.

12.6 The Rise of Ripple-Aligned Technology

For over a century, our devices — from the lightbulb to the supercomputer — have been designed around visible output and measurable flow. They operate by forcing electrical charge, switching voltage levels, or toggling binary states.

This approach worked well in a world where energy was thought to flow like a substance and where information was reduced to pulses and bits.

But as we now understand through Universal Energy Dynamics (UED), the universe doesn't work by flow. It works by resolution. Everything from light to logic is a ripple alignment phenomenon — a field condition reaching the right threshold for coherent transition.

As this new understanding takes hold, technology is about to shift — radically.

From Charge to Coherence

In the UED framework, the future of technology is not about:

- Pushing more current.
- Reducing component size.
- Stacking more transistors.

It's about mastering ripple compatibility.

Devices will no longer treat electrons as carriers and voltages as triggers. Instead, they will:

- Resonate with ambient field conditions.
- Respond to ripple configurations that match their internal tuning.
- Function only when ripple permission is granted, based on ATT (Alignment Threshold Time) windows.

This marks the rise of ripple-aligned technology — where precision is not defined by tolerance in voltage or timing jitter, but by harmonic compatibility with the structured vacuum itself.

Core Design Parameters for Ripple-Tech

Here's what future engineers and physicists will focus on:

1. Ripple Frequency Compatibility

Every interaction — whether light, sound, or data — is a ripple disturbance in the field. Ripple-aligned devices must:

- Match the input frequency with the material's natural resonance.
- Prevent interference by tuning out of band ripple paths.
- Adapt in real-time to field drift and environmental ripple strain.

Imagine memory chips that don't store binary states, but capture ripple frequencies like a tuning fork — and hold them in locked alignment.

2. Field Stability Windows

Materials will be graded not by hardness or conductivity, but by how stable their internal field geometry remains under environmental ripple variation.

This redefines what we mean by:

- Durability → becomes ripple resilience.
- Efficiency → becomes coherence retention.
- Noise immunity → becomes internal field shielding.

You won't need more shielding. You'll need field-stable substrates that naturally align with the background SVE (Structured Vacuum Energy) grid.

3. ATT Gating and Resonance Capture

Current circuits use timing pulses (clocks) to coordinate computation. Ripple-aligned systems will use ATT gates — controlled windows where ripple resolution is allowed under exact threshold conditions.

These systems won't "tick" — they'll flow only when harmony exists.

Imagine:

- Ripple sensors that capture field variations without ever touching the object.
- Logic gates that open not on voltage, but when field tension matches a known pattern.
- Ripple memory that remembers alignment states, not stored charges.

This moves us away from brute-force computation to elegant ripple negotiation — interacting with the universe on its terms, not ours.

From Quantum Computing to Ripple Intelligence

Quantum computers, as currently imagined, rely on fragile superpositions and entanglement — easily disturbed, hard to scale. But under UED, a more reliable approach emerges:

- Not simulating superposition, but maintaining ripple identity across aligned nodes.
- Not probabilistic collapse, but ATT-based decision points triggered by ripple field evolution.

This opens the door to:

- Ripple-based processors that operate on logic of permission, not voltage.
- Ripple AI that interprets alignment patterns in real time

and reacts with natural field behavior.
- Distributed sensing networks that feel ripple tension shifts across space — like a nervous system for the Earth.

We're not building machines. We're designing instruments to listen and respond to the field.

Redefining Precision

In ripple-aligned tech, precision is no longer:

- Nanometers of etching.
- Gigahertz of switching speed.
- Bits per second.

Instead, precision becomes:

"How perfectly does this device allow ripple resolution with minimal distortion?"

This is subtle technology — invisible, efficient, calm.

It doesn't brute-force answers. It resonates with them.

12.7 Closing Insight – Precision Is Permission

"Quantum isn't small. It's aligned."

For decades, we believed that progress in physics and technology meant diving deeper into the small — the subatomic, the nano, the quantum. We celebrated how small we could make the transistor, how short a pulse we could emit, how little energy we could use to flip a bit.

But the real breakthrough wasn't in the scale. It was in the synchronization.

What we called "quantum" was never about size. It was about selectivity — the universe saying:

"Now, the conditions are just right. You may resolve."

That is not a property of particles. It is a property of permission.

Precision Is Not Control — It's Compatibility

All our greatest technological feats — from the laser to the transistor to the superconducting magnet — work not because we dominate matter, but because we match its ripple behavior with the surrounding field.

They are not perfect because they're powerful. They are perfect because they fit the field's logic without resistance.

And that is the true meaning of precision: Not accuracy to our instruments, but alignment to the universe.

The Path Forward: Deeper Resonance, Not Deeper Cuts

As we move forward, the temptation will be to push harder:

- Smaller transistors.
- Faster processors.
- More data.

But UED invites a different path:

- Can we build systems that only act when alignment allows it?
- Can we tune to the ripple rather than shove energy into resistance?
- Can we measure success not in output, but in how silently we resolve tension?

We must begin to see our inventions not as tools of extraction, but as agents of coherence.

From Quantum to Ripple Intelligence

The future is not quantum computing. It's ripple intelligence. Not about probabilistic states, but about field-aware design that listens, waits, and acts only when the field says yes.

That future is quieter. More efficient. More intelligent. And profoundly more aligned with the natural order.

We never needed to conquer the atom. We only needed to understand the ripple's permission.

Precision is not a matter of scale. It is a matter of respect — for the field, for the threshold, for the harmony waiting to unfold.

And once we align with it...Everything flows — not because we forced it to, but because the universe finally said: *"Now, you may."*

Interlude – We Never Controlled Energy

We Aligned with It.

We thought we were masters of power.

From the atom split in fury, to the laser beam whispering through fiber, to the silent turning of wind turbines — we've built an empire on the illusion that energy is ours to command, transfer, store, and extract.

But every so-called "source" turned out to be a condition. Every output we celebrated was just a resolution. And every machine we praised was simply a field interface, built to allow the universe to do what it was already trying to do — but with less resistance.

Whether in the core of a nuclear reactor or the layered lattice of a solar cell, we were never harnessing power. We were permitting harmony.

All of Part IV revealed this deeper truth:

- Nuclear explosions are not the triumph of mass-energy conversion, but the violent collapse of ripple coherence.
- Renewables don't "generate" — they mediate realignment.
- Quantum technologies don't defy logic — they expose the hidden elegance of ripple thresholds.

In each case, progress did not come from force. It came from fidelity — our growing ability to sense, respond to, and tune into the field itself.

We are no longer engineers of output. We are becoming curators of alignment.

And so, we arrive at the next frontier — perhaps the most intimate of all: Not the machines we build, But the intelligence that arises when field and form meet.

Not the energy we manage, But the awareness that emerges when resolution becomes conscious.

If everything we called energy was simply ripple discomfort seeking rest, Then what is life?

What is mind?

What is the observer, who stands at the edge of every collapse and says: *Now.*

Part V: Implications and Applications

We never taught energy. We taught metaphors to cope with phenomena we didn't fully understand.

Chapter 13: Educational Reset

"We didn't teach energy. We taught equations with missing meaning."

Open any school physics textbook, and you'll find energy treated as a quantified substance. Defined in joules. Tracked in bar charts. Plugged into equations like currency in a ledger.

The standard pedagogical model goes like this:

- Energy is conserved.
- It can be stored, transferred, or transformed.
- It exists in various "forms" — kinetic, potential, thermal, chemical.

Students are taught to count it, but not to understand it.

What goes missing is the why — why does energy behave this way? What is it *really*? It is presented as something you have, not a condition that arises.

This misunderstanding is not just academic — it shapes how entire generations think about power, effort, value, and even life itself.

13.1 How We Ended Up Teaching a Lie

Textbooks rarely question their own metaphors. They still depict:

- Flowing charges.

- Energy "stored" in springs and fuels.
- Light as particles or waves moving from one point to another.

These simplifications were meant to be stepping stones, but they became permanent scaffolding. The result? An education system that teaches energy as a commodity rather than a ripple-resolution event.

We taught motion but not misalignment. We taught formulas but not field behavior. We taught power as output, never as permission granted by the universe.

13.2 Dismantling the "Energy Unit" Myth

Under Universal Energy Dynamics (UED), we no longer see energy as a stored asset. We see it as a local consequence of ripple discomfort resolving at the observer's node.

This changes everything we teach:

Old Paradigm	UED Paradigm
Energy is something	Energy is a result
It flows	It resolves
It can be stored	Misalignment can persist
Energy is conserved	Field tension redistributes
Energy causes change	Field permission enables change

We must shift from accounting to awareness. From "How much?" to "Where and why does this align?"

13.3 Teaching Ripple Behavior from the Start

Imagine a physics curriculum that begins not with:

"An object has potential energy." But with: *"This system holds ripple tension — and here's how it can resolve."*

Students would learn:

- That motion is not caused, it is permitted.
- That fields are not passive — they are actively strained.
- That heat is ripple turbulence, not a form of energy.
- That mass is not "converted," but reconfigured tension.

Simple experiments like dropping a ball or lighting a bulb can be reframed as field reconfiguration events:

- The falling ball is not releasing energy. It is seeking ripple relaxation.
- The bulb doesn't "get energy." It resolves local discomfort induced by a distant oscillating source.

This approach also invites interdisciplinary learning:

- Music as ripple patterning.
- Art as visual field resonance.
- Biology as alignment behavior within living systems.

Science would no longer be taught as dead content, but as living ripple dynamics — with the observer always in the loop.

13.4 Rewriting the Textbook Narrative

To realign education with UED:

- Energy chapters must be rewritten as field behavior chapters.
- Thermodynamics must begin with ripple tension, not heat and work.
- Electromagnetism must be reframed as ripple modulation, not charge in motion.
- Quantum theory must shed its probabilistic mystery and

embrace ATT and ripple permission.

And most importantly, we must give students the power to ask: *"What is resolving here? And why now?"*

Instead of memorizing outcomes, they would begin sensing the language of the field — and seeing every phenomenon as a negotiation of ripple alignment.

13.5 The Teacher's Role in a Ripple Age

In this new model, the teacher is no longer a provider of facts, but a guide through misalignment.

They help students:

- Recognize strain in their thinking.
- Align concepts across subjects.
- Become comfortable with tension — intellectual, emotional, and scientific.

Teaching becomes less about delivering knowledge and more about facilitating resonance.

Because the deepest learning doesn't happen when facts are given. It happens when the mind says: *"This finally aligns."*

13.6 Closing Insight – What We Really Teach

We never taught energy. We taught metaphors to cope with phenomena we didn't fully understand.

Now, with the language of ripples, tension, alignment, and resolution, we have the chance to teach reality as it is — alive, interconnected, elegant.

The classroom of the future will not revolve around equations. It will revolve around permission — the moment when a mind aligns with the field... and *knows*.

Chapter 14: Policy, Climate, and the Myth of Energy Crisis

"The planet never ran out of energy. It ran out of alignment."

Every few decades, we're told the same story in new packaging:

- "We're running out of energy."
- "Fossil fuels are peaking."
- "Climate catastrophe is energy misuse."
- "Renewables must replace everything — urgently."

From oil embargoes to carbon budgets to net-zero campaigns, the dominant narrative has always framed the problem as resource depletion — as if the planet is a barrel that's nearly empty.

But this view only makes sense if energy is a substance — something that can be extracted, consumed, and exhausted.

It isn't.

Under Universal Energy Dynamics (UED), energy is not a fuel, a flow, or a stored quantity. It is a consequence — of ripple discomfort resolving in local fields. And the real crisis isn't energy shortage — it's ripple misalignment.

14.1 The Earth as a Ripple System

The Earth is not an energy tank. It is a living field equilibrium.

- The atmosphere = a dynamic ripple buffer.
- The oceans = fluid ripple stabilizers.
- The crust = slow-release tension modulation.
- The biosphere = self-regulating ripple feedback systems.

When this balance holds, ecosystems flourish. When it's distorted, turbulence increases — not just in weather patterns, but in social systems, economic structures, and emotional fields.

Climate change is not just heat. It is persistent misalignment in the Earth's ripple balance, amplified by:

- Excessive material disruption.
- Artificial field geometries (urban heat zones, EM interference).
- Technological overreach without ripple awareness.

The crisis isn't fossil fuel use per se. It's fossilized thinking — designing systems that push ripple tension into the planet without allowing proper resolution.

14.2 Rethinking Sustainability: From Resource to Resonance

"Sustainability" is often framed in metrics:

- Carbon output.
- Renewable input.
- Efficiency ratios.
- Circular economies.

But UED reveals that what matters is not the quantity used, but the quality of ripple harmony maintained.

Let's compare:

Conventional View	UED View
Reduce emissions	Reduce ripple strain
Use clean energy	Design clean resolution paths
Switch fuels	Switch field interaction models
Limit consumption	Enhance ripple coherence

True sustainability isn't about "less." It's about better ripple conduct — doing more with minimal field distortion.

A machine that burns nothing but distorts the field violently (e.g., through vibrations, EM bursts, noise) is less sustainable than one that uses fossil fuel but resolves ripple gently and cleanly.

14.3 Policy Without Physics Is Blind

Most energy and climate policy today is material-focused, not field-aware:

- Taxing emissions without questioning the underlying field disturbance.
- Incentivizing renewables that are ripple-incompatible with the local environment.
- Pushing global targets without acknowledging regional field dynamics (e.g., sun-angle inefficiency in some geographies, or wind chaos near sensitive habitats).

Policies must evolve from:

"How much are we using?" to "What strain are we introducing — and how is it being resolved?"

This opens up powerful new tools:

- Ripple Impact Assessments alongside Environmental Impact Assessments.
- Field Harmony Indexes for cities, machines, and industries.
- Policy triggers based on real-time field turbulence metrics,

not just emissions reports.

Governance becomes not about restriction, but about alignment stewardship.

14.4 The Myth of Transition

We often hear of a global energy transition — from fossil to clean, from centralized to distributed, from polluting to sustainable.

But under UED, the transition is deeper:

We are not moving from dirty to clean. We are moving from ignorance to intelligence.

From extracting energy to aligning ripples. From forcing outcomes to facilitating resolution.

That's not a switch. It's a civilizational realignment.

It doesn't just change machines. It changes how we define success, progress, cost, and even crisis.

14.5 Closing Insight – The Planet Doesn't Need Saving. It Needs Harmonizing.

We keep trying to save the planet with new machines, new metrics, and new markets.

But Earth was never broken. It was only misread.

The energy crisis was a crisis of perception — mistaking field tension for fuel shortage, mistaking consumption for necessity,

mistaking entropy for end.

The planet isn't asking us to stop using energy. It's asking us to stop forcing unresolved ripple strain into its systems.

The future of policy is not decarbonization. It is re-alignment — designing everything we do to reduce ripple conflict and enhance field resolution.

From cities to circuits, homes to highways, supply chains to schoolbooks...

It's not how much energy we consume that defines our fate. It's how much alignment we leave behind.

Chapter 15: Toward a Ripple-Based Civilization

"The next revolution won't be industrial. It will be align-mental."

Every past civilization can be measured by what it extracted:

- Stone Age: rocks and minerals.
- Bronze and Iron Ages: alloys and tools.
- Industrial Age: coal, steam, machines.
- Digital Age: data, attention, electrons.

Our tools advanced. Our world transformed. But the mindset remained:

"Take more. Burn faster. Store better. Repeat."

Even in sustainability efforts, the underlying model is extraction:

- Extract energy "cleanly."
- Extract efficiency from systems.
- Extract output from nature with minimal waste.

But Universal Energy Dynamics (UED) reveals the flaw: We've been designing around outputs — instead of designing for alignment.

The next step for humanity is not better extraction. It is field-integrated design — systems, societies, and selves that operate in ripple coherence with their environment.

We won't win by overpowering nature. We'll thrive by finally participating in its rhythm.

15.1 What a Ripple-Based Civilization Looks Like

Let's imagine a society that no longer asks:

- "How much can we produce?"
- "How efficiently can we consume?"
- "How big is our energy supply?"

And instead asks:

- "How cleanly can we resolve tension?"
- "How softly can we operate?"
- "How aligned are our lives with the field we live in?"

In such a civilization:

Homes

- Are not insulated boxes, but ripple modulators — balancing internal and external field strains to maintain comfort without mechanical input.

Transportation

- Becomes a matter of minimal field disruption, using paths and vehicles designed not just for speed or fuel economy, but for coherence with land, weather, and biological systems.

Architecture

- Shifts from vanity monuments to field-resonant structures — designed to minimize internal turbulence and enhance local alignment.

Healthcare

- Moves from symptom management to bio-ripple diagnostics — detecting internal misalignments before they become material disease.

Energy systems

- Are not grids distributing power, but ripple webworks — dynamically aligning tension across community-scale fields.

In short: We move from engineering consumption to engineering consonance.

15.2 The Industries of Tomorrow

New fields will emerge that make no sense under the old paradigm, but become obvious under UED:

1. Ripple Diagnostics

Tools and methods to detect:

- Field misalignment in ecosystems.
- Ripple turbulence in machines.
- Pre-collapse ATT zones in social systems or supply chains.

2. Alignment Engineering

Designing machines and environments that:

- Require minimal ripple effort to function.
- Use local field permission instead of brute-force input.
- Self-correct misalignment via ripple feedback.

3. Bio-Alignment Technologies
Working with the body, not against it:

- Materials that synchronize with biological fields for faster healing.
- Neural systems that detect thought misalignment as stress signatures.
- Adaptive clothing and spaces that maintain human ripple coherence during work or sleep.

This is not utopia. It is field literacy made real.

15.3 The Economic Shift – From Growth to Grace

A ripple-based civilization doesn't chase exponential GDP. It chases field stillness — the kind that allows complexity to thrive without collapse.

Economic systems will reward:

- Designs that resolve tension rather than redirect it.
- Processes that align faster, not just scale wider.
- Systems that listen to feedback, not override it.

Efficiency will no longer be a ratio of input to output. It will be a measure of how gently and quickly alignment is achieved.

15.4 Human Alignment – The Final Frontier

Ultimately, the ripple civilization is not about technology. It is about intelligence — not in circuits, but in selves.

As individuals:

- We stop forcing life into routines that break our rhythms.
- We stop valuing stress as productivity.
- We stop mistaking friction for ambition.

The aligned human:

- Learns when to act — and when to yield.
- Listens more than pushes.
- Lives in a way that doesn't just reduce damage — it enhances resolution for others.

This is not passivity. It is mastery through participation.

We become agents not of power, but of permission — living beings whose very presence helps the field settle.

15.5 Closing Insight – Alignment Is the Highest Civilization

The next civilization will not be measured by energy consumed, land conquered, or wealth generated.

It will be measured by how much ripple strain it prevents, how much alignment it restores, and how little force it requires to thrive.

We will build less. And everything we build will matter more.

We will consume less. And everything we touch will carry coherence.

We will speak less. And everything we say will align someone.

Because ultimately:

Civilization is not what we construct. It is the quality of permission we give to the universe through our existence.

And in that stillness — in that deep, collective harmony — we finally become intelligent.

Epilogue – The Field Remembers

We came searching for energy. We found alignment.

We questioned machines. We ended up questioning ourselves.

What began as a journey through wires, fuels, atoms, and waves became something deeper — a revelation that nothing flows, nothing moves, nothing burns, unless the field allows it. Unless the ripple tension finds its path to resolve.

In every explosion we feared, in every bulb we lit, in every storm we tried to tame, the same truth echoed back:

Energy was never a thing. It was always a sign. A sign that something strained — and something realigned.

We were never separate from this story. We were the observer that shaped it. We were the system being resolved.

The sun doesn't shine to power us. The earth doesn't spin to serve us. They simply are — in coherent ripple motion.

And the wisest thing we can do is match that rhythm, not fight it.

In classrooms, we will now teach ripple resonance before we teach resistance. In labs, we will measure permission before power. In policies, we will model harmony before efficiency.

In minds, we will awaken stillness before ambition.

Because we've seen the alternative.

We've seen what happens when alignment is ignored — when the field is forced, the environment destabilized, the mind overclocked, and the machine glorified.

We called it progress. But it was only acceleration without permission.

Now, we pause. We listen. We align.

This is not the end of the energy story. It is the beginning of a more intelligent civilization — not driven by hunger for power, but by awareness of tension and grace in its resolution.

In this new world:

- The best designs leave no trace but harmony.
- The greatest minds quiet the noise around them.
- And the most advanced systems are the ones that barely need to act — because everything around them is already aligned.

So if you must remember one thing from this book, remember this:

Energy is not what moves us. Alignment is what allows us to move — meaningfully. And in a world of noise, force, and urgency... The most powerful thing you can do is listen deeply to the ripple beneath all things.

Because the field remembers. And it's waiting for you to align.

Appendices

"Before we understand the field, we must speak its language."

Appendix A – UED Equations

(with Real-Life Interpretations)

"Equations are not truths. They are agreements with the field."

The Universal Energy Dynamics (UED) framework is not built on force, flow, or conserved quantities. It is built on ripple behavior, alignment conditions, and the field's willingness to resolve discomfort.

As such, the equations of UED don't describe *what* moves, but *why* movement happens at all. They don't predict absolute outcomes — they define conditions for permitted resolution.

This appendix outlines key UED relationships, each followed by a real-life interpretation to bridge theory and experience.

A.1 The Ripple Resolution Principle

$$E = R_{\text{eff}} \cdot \Delta\Psi$$

Where:

- E = Observed energy event (localized ripple resolution)
- R_{eff} = Effective ripple resistance (field tension at observation point)
- $\Delta\Upsilon$ = Ripple misalignment magnitude across the system

Interpretation:

What we call "energy" is not something that flows — it's the *visible effect* of the field realigning under strain. The greater the misalignment, and the more resistant the environment, the more pronounced the observable consequence.

Example: A lightning strike isn't a transfer of stored charge. It's a sudden collapse of ripple misalignment in a highly strained atmospheric field.

A.2 Alignment Threshold Time (ATT)

$$\text{ATT} = \frac{C_{\text{sys}}}{R_{\text{env}}}$$

Where:

- ATT = Minimum time required for ripple alignment to be permitted
- C_{sys} = Coherence of the system attempting alignment
- R_{env} = Ripple resistance in the surrounding field

Interpretation:

Every event — from a nuclear trigger to a neuron firing — requires a window of allowed alignment. If coherence is too low or environmental resistance too high, the resolution cannot proceed.

Example: Fusion requires extreme ATT collapse — achieved only under immense field compression and heat in stars or reactors. Without sufficient ripple readiness, no fusion occurs — no matter how much fuel is present.

A.3 Ripple Density Gradient (RDG)

$$\nabla \rho_r = \frac{d(\Psi)}{dV}$$

Where:

- ρ_r = Ripple tension density
- Ψ = Misalignment potential
- V = Spatial volume

Interpretation:

Ripple gradients determine not flow, but where discomfort is likely to resolve. Steeper gradients mean higher likelihood of resolution events — heat, current, movement, or breakdown.

Example: Wind is not air flowing due to pressure. It is the macro ripple behavior of atmospheric fields adjusting to large-scale misalignment.

A.4 Ripple Reversion Delay (RRD)

$$t_d = \tau_{\text{field}} \cdot \left(1 + \frac{D_{\text{sys}}}{C_{\text{env}}}\right)$$

Where:

- t_d = Time delay for ripple tension to return to equilibrium
- τ_{field} = Natural damping constant of the local field
- D_{sys} = Disruption magnitude
- C_{env} = Environmental coherence (ripple-damping capacity)

Interpretation:

The time it takes for any system to "settle down" is not just about mechanics — it's about how coherent the field is in absorbing and distributing ripple strain.

Example: Emotional tension in a group, or "bad vibes" in a room, follow the same model. High disruption + low coherence = longer misalignment and delayed calm.

A.5 No-Flow Constraint (Fundamental UED Principle)

$$\text{Energy}_{\text{unresolved}} \equiv \text{Ripple}_{\text{misaligned}}$$
$$\text{Flow} = 0 \quad (\text{Always})$$

Interpretation:

There is no such thing as energy flowing through space or wires. What appears as flow is a field-wide ripple rearrangement, perceived as sequence — not transmission.

Example: Electricity doesn't travel down a wire like water in a pipe. The field trembles — and each point locally resolves based on ripple tension. What we call current is just synchronized discomfort resolution.

A.6 The Stillness Law (Field Ground State)

$$\sum \vec{\tau} = 0 \quad (\text{Ideal ripple-neutral system})$$

Where:

- $\vec{\tau}$ = Local ripple strain vectors
- The sum being zero means no net discomfort exists in the field — perfect coherence.

Interpretation:

The universe doesn't "prefer" motion or energy. It seeks stillness — not absence of activity, but absence of unresolved tension.

Example: A perfectly tuned musical note resonates effortlessly, needing no extra force to sustain. That's what stillness feels like — everything aligned, nothing resisting.

Final Thought: Equations Are Not the Truth. They Are the Invitation.

Every formula in UED is a tool to sense alignment, not predict outcomes like a machine. Unlike classical physics, where math dominates meaning, UED treats equations as maps of ripple permission.

You are never calculating energy. You are tracing where tension can resolve — and under what conditions.

In this way, the universe becomes not an engine to measure, but a conversation to participate in.

Appendix B – Glossary of Core UED Terms

"Before we understand the field, we must speak its language."

Ripple

Definition: A ripple is a localized disturbance in the structured vacuum field, representing a deviation from perfect stillness. It is not a wave in a medium but a tension configuration seeking resolution.

Clarification:

- In UED, ripples are not "carriers" of energy.
- They are field imbalances — and only when resolved do we perceive energy, motion, or light.

Everyday Analogy: A ripple is like the subtle strain in a stretched bedsheet — barely visible, but it shifts everything around it. When this ripple aligns or collapses, something *happens* — a flash, a sound, a thought.

Energy is not what travels. The ripple is what happens before the universe permits something to be felt.

ATT – Alignment Threshold Time

Definition: ATT is the minimum time window required for a ripple to realign (resolve) within its field environment.

Clarification:

- Every field has a limit to how quickly or slowly it can permit change.
- ATT defines that boundary.
- If a ripple tries to resolve before the environment is ready — it fails, rebounds, or delays.

Why It Matters:

- It explains why some chemical reactions need catalysts.
- Why fusion happens only under certain extreme conditions.
- Why ideas don't land until the listener is "ready."

Real-World Example: In social dynamics, ATT is the reason why some truths take time to be accepted — not because they're false, but because the field isn't ready to permit their alignment.

Events don't happen because we trigger them. They happen when ATT permits resolution.

Tension (Field Tension / Ripple Tension)

Definition: Tension in UED is the strain within a field due to ripple misalignment. It is the source of all observable events — not because it causes motion, but because it demands resolution.

Clarification:

- Heat is unresolved tension at the micro level.
- Electricity is guided tension resolution.
- Pressure, light, sound — all are ripple tension interactions.

Misconception Addressed: Tension is not force. It is the cause of the condition we interpret as force. It is the pre-condition of energy.

Example: In a stretched rubber band, we see tension. In UED, the entire universe is made of such subtle tensions — some visible, most invisible — all waiting for alignment.

Where tension exists, resolution is inevitable — unless something blocks it.

Polarity

Definition: Polarity is the orientation of ripple behavior relative to surrounding field geometry — it dictates the direction of resolution and interaction.

Clarification:

- It is not charge-based (like positive/negative in electromagnetism).
- UED polarity describes alignment preference — which way the ripple wants to collapse or connect.

Why It's Important:

- Opposing polarities invite resolution (like compression and expansion in sound).
- Identical polarities resist each other — causing persistent tension.

In Systems:

- **Mental polarities:** Optimism vs fear — ripple directions in thought space.
- **Biological polarities:** Left/right neural hemispheres creating ripple complementarity.

Cosmic View: All cosmic evolution is a game of polarity balancing — from the spin of galaxies to the thoughts in your head.

Polarity is not conflict. It is the geometry of potential harmony.

Closing Insight: The Glossary Is the Gateway

UED is not a new branch of physics. It is a new lens for reality, with its own vocabulary — one that honors motionless cause, not just observable consequence.

When you speak in ripples, tensions, permissions, and ATT, you stop asking:

What happened? And start asking: *What was allowed to happen — and why now?*

Appendix C – UED vs Classical Energy Models

"We didn't evolve beyond classical physics. We evolved beyond its illusion."

This comparison outlines the fundamental differences between Universal Energy Dynamics (UED) and classical energy models, including Newtonian mechanics, thermodynamics, and mainstream electromagnetism and quantum theory.

The shift isn't just in terminology — it is a complete reversal of causality and ontology:

- Classical models see energy as a thing.
- UED sees energy as a ripple resolution event — a consequence, not a quantity.

Let's map the contrast clearly.

C.1 Ontology: What Is Energy?

Classical View	UED View
Energy is a measurable quantity that exists in matter or systems.	Energy is not an object. It is the local effect of ripple tension resolving in a structured vacuum field.
It can be stored, transferred, or transformed.	It cannot be stored or transferred — only permitted as alignment occurs.
Energy is conserved in all closed systems.	What is conserved is ripple equilibrium — not "energy." Conservation is a byproduct of field balance, not a rule.

C.2 Motion and Work

Classical View	UED View
Objects move when force is applied; work is force × distance.	Motion occurs only when ripple permission is granted — force is not a cause but a response to unresolved ripple strain.
Work requires energy input; energy is lost as heat.	There is no input/output — only ripple realignment. Heat is unresolved turbulence, not energy "loss."

C.3 Thermodynamics

Classical View	UED View
Heat is energy in transit due to temperature difference.	Heat is ripple noise — random unresolved micro-strain. Temperature is not cause, but a symptom of ripple turbulence.
Entropy increases over time in closed systems.	Entropy is not a law — it's a field misalignment tendency when ripple coherence is lost. Systems can regain alignment.
Energy moves from high to low potential.	Resolution occurs where ripple strain finds lower resistance — not necessarily spatial or thermal gradients.

C.4 Electricity and Magnetism

Classical View	UED View
Electrons move through conductors, generating electric current.	No flow occurs — ripple tension realigns locally across the field. "Current" is synchronized ripple resolution.
Magnetic fields arise from moving charges.	Magnetism is a field spin behavior — ripple loops, not charge spirals. Polarity arises from ripple vector orientation.
Electric potential is due to charge difference.	Potential is ripple strain. Voltage = ripple pressure, not stored charge.

C.5 Quantum Theory

Classical/Quantum View	UED View
Photons are energy packets.	Light is a ripple resonance event, not a particle. No packets, just frequency-specific alignment collapses.
Particles have wave-particle duality.	No duality — everything is ripple configuration. "Particle" and "wave" are observer-conditioned metaphors.
Quantum events are probabilistic and random.	Apparent randomness = ATT barrier fluctuations. Collapse happens only when alignment threshold is crossed.
Entanglement involves instant communication.	Entanglement is shared ripple identity — no signal, just coherent field relationship.

C.6 Nuclear Reactions

Mainstream View	UED View
Mass is converted into energy ($E = mc^2$).	Nothing is converted — ripple disintegration releases strain. $E = mc^2$ is an outcome, not a cause.
Binding energy is the energy required to split a nucleus.	Binding energy = ripple stability status. "Energy release" is field realignment, not stored energy escaping.
Chain reactions are driven by neutron emissions.	Chain reactions are ATT cascade events — triggered realignments, not physical dominoes.

C.7 Renewable Energy

Conventional View	UED View
Solar panels absorb photon energy.	Panels are resonant receivers of ripple frequency, not photon catchers. Electron movement = ripple tension resolution.
Wind turbines convert kinetic energy into electricity.	Wind is ripple strain — turbines are field tension translators, not energy extractors.
Hydro power is gravitational energy being harnessed.	Water fall due to ripple descent — gravity is vertical ripple compression, not a pulling force.

C.8 Final Perspective

Classical models helped us build engines, machines, and theories — but they boxed reality into discrete quantities, flows, and packets. They confused metaphor for mechanism.

UED reframes the universe as a stillness-seeking field where all phenomena are ripple permissions — not power plays.

We're not here to reject classical science, but to liberate reality from its misinterpretations — to restore the causal primacy of alignment, not motion.

The revolution is not in new math. It's in remembering that the universe never moved — it only resolved. And we finally learned how to watch it with the right questions.

Appendix D – Suggested Classroom Experiments

"Don't measure energy. Witness alignment."

UED (Universal Energy Dynamics) doesn't ask classrooms to throw out thermometers or voltmeters — it asks them to reinterpret what they're observing.

This section offers classroom-friendly experiments (using existing tools and simple setups) to demonstrate ripple behavior, ATT dynamics, and the illusion of flow. Each experiment is followed by a UED interpretation, so students don't just perform activities — they learn to see differently.

D.1 The Ripple Path of Light – Refraction Without Flow

Setup:

- A transparent glass of water.
- A straight straw or pencil.
- A light source.

Procedure:

- Place the straw diagonally in the glass of water.
- Observe how the straw appears "bent."
- Move the light source around and observe brightness patterns.

Conventional Interpretation: Refraction due to change in light speed across media.

UED Interpretation: The field tension across air and water is different. Light is not "slowed down." The ripple permission geometry changes at the boundary. What you observe is not a "bending" of something flowing, but a new local alignment path for ripple resonance.

Light didn't bend — the field re-permitted it.

D.2 Electric Circuit with Delay – Observing ATT in Action

Setup:

- A long wire (~10 meters coiled).
- A battery or power supply.
- A light bulb or LED.
- A switch.

Procedure:

- Build a simple circuit with the long wire and bulb.
- Flip the switch and observe the bulb's response.
- Try adding a capacitor or resistor to compare timing.

Conventional Interpretation: Current flows through the wire; small resistance affects response time.

UED Interpretation: There's no flow. The bulb lights up when ripple tension across the circuit aligns and crosses its local ATT. Delays arise not from travel, but from field preparation — just like someone clearing their throat before speaking.

You didn't send electricity. You allowed permission for light to resolve.

D.3 Boiling Water with and Without a Lid – Turbulence as Misalignment

Setup:

- Two identical beakers of water.
- One lid or cover.
- Heat source (hot plate).

Procedure:

- Heat both beakers simultaneously, one with a lid, one open.
- Observe the time it takes to boil, steam behavior, and noise.

Conventional Interpretation: Covered water heats faster due to heat retention.

UED Interpretation: Boiling is ripple turbulence becoming strong enough to exceed ATT. The lid reduces dissipation, allowing internal ripple coherence to build up more quickly. The open beaker disperses strain externally — delaying resolution.

The field didn't heat faster. It misaligned slower under the lid.

D.4 The Levitating Magnet – Ripple Rejection in Coherence

Setup:

- A strong neodymium magnet.
- A superconducting disk cooled with liquid nitrogen (if available).
- Safety equipment.

Procedure:

- Cool the superconductor.
- Place the magnet gently above the disk.
- Observe levitation.

Conventional Interpretation: Meissner effect and magnetic field expulsion.

UED Interpretation: The superconductor achieves near-perfect ripple alignment. It no longer permits internal ripple strain. The magnet's presence introduces incompatible tension — which the field rejects, resulting in levitation.

What rises is not the magnet. What pushes is coherence refusing distortion.

D.5 Group Clapping Exercise – Human ATT Visualization

Setup:

- A group of students.
- A timer or rhythm metronome.

Procedure:

- Have students randomly clap once per second.
- Now ask them to all clap together on a signal.
- Finally, ask them to maintain synchronized clapping without signal.

UED Interpretation: Synchronization doesn't require command — it requires shared permission. ATT in the group must be respected. The exercise shows how alignment emerges or collapses when thresholds of internal coherence are breached.

Rhythm is a group ripple. Collapse happens when someone claps before ATT permits.

D.6 Spinning Fan in a Closed Box – Misalignment Storage

Setup:

- A small table fan.
- A cardboard box that fits over it.
- Thermometer (optional).

Procedure:

- Turn the fan on, then cover it with the box.
- Observe temperature rise, sound, and feel.
- Remove the box and notice the immediate "whoosh."

UED Interpretation: The box prevents ripple dissipation. Turbulence accumulates. The moment it's lifted, stored misalignment is released rapidly. There's no energy stored — just tension waiting for permission to escape.

Closing Thought: Teaching the Field

These experiments aren't designed to prove UED. They're meant to untrain the eye, to help students ask deeper questions:

- What am I really observing?
- Is anything actually moving?
- Or is the field simply resolving?

The best experiment is always the one where the student doesn't just measure — but feels the ripple shift in their own awareness.

Because at the heart of science is not data. It is alignment with what's real — even when it doesn't flow.

Acknowledgments

This book was never just written — it was allowed to happen.

It emerged not from effort alone, but from a quiet field of conversations, contradictions, long walks, and moments of dissonance that eventually chose to align.

To the silent ripples that first whispered the illusion of "energy" into question — thank you. You began this inquiry long before it had a name.

To every student who ever asked *"But where does the energy really go?"* — you were right to ask. This book is your answer and your invitation.

To the thinkers, physicists, philosophers, and teachers — those who agreed, disagreed, debated, or dismissed — your responses helped shape the clarity and courage behind these pages. Your resistance was also a ripple. It mattered.

To those rare souls who listen deeply — who see beyond wires and numbers, and sense the stillness behind every movement — you are the real scientists of the future.

To the close friends and readers who believed in this project before it became structured: your feedback, questions, and patience helped anchor the chaos into coherence.

To the field itself — for permitting this work to resolve — my deepest gratitude. Without your silent permission, no alignment would have happened. No word would have landed.

And finally, to the Observer — not just the reader, but the intelligence behind the reader — thank you for completing the ripple.

This book ends with you.

— *Sandeep Chavan*

About the Author

Sandeep J. Chavan is an independent physicist, educator, and philosopher best known for developing the **Universal Energy Dynamics (UED)** framework — a field-first approach that challenges the foundational assumptions of classical and quantum physics.

With a background in engineering and decades of experience teaching physics at the grassroots level, Chavan bridges the gap between scientific rigor and intuitive understanding. He writes not to impress institutions, but to empower individuals — students, thinkers, and seekers — to see reality beyond inherited equations.

Chavan has authored over 100 books spanning science, philosophy, psychology, and consciousness. His work questions not just how the universe works, but why we believe the stories we tell about it.

He lives a simple life, disconnected from mainstream academia, yet deeply connected to the ripples that shape existence.

This book is part of that ripple.

Bold Declarations of Universal Energy Dynamics (UED)

"We do not fix physics. We dissolve its illusions."

There is no such thing as energy.

What we call "energy" is not a substance, object, or unit. It is the observable consequence of ripple tension resolving in a field. Nothing flows. Nothing moves. Everything aligns — or waits.

Electricity does not flow through wires.

There is no current traveling. No electrons marching. What we observe as electric "power" is a synchronized ripple realignment, permitted across a strained field.

Light is not a particle or wave. It is a permission.

A photon is a field event — not a thing. When a ripple aligns with a receiver's ATT (Alignment Threshold Time), we see light. Before that moment, nothing exists. Not even the "photon."

Mass is not converted to energy. Ever.

$E = mc^2$ is a brilliant consequence — but not a cause. What nuclear reactions release is stored ripple tension, not "mass turned into energy." Matter is not a battery. It is a configuration waiting to misalign.

Gravity is not a force.

There is no attraction. No pull. Gravity is the compression geometry of ripples in a structured vacuum field, seeking equilibrium.

There are no fields "of" something — only the field.

Electromagnetic, gravitational, thermal — these are not separate forces. They are ripple behaviors of one field in various alignment states. UED is unification by coherence, not by equations.

Resistance is not opposition. It is misalignment.

Electrical resistance, friction, stress — all are ripple tensions that the field hasn't resolved.

Superconductivity = perfect alignment.

Efficiency = minimal ripple noise. Innovation = better permission.

You are a ripple in the field — not separate from it.

Your thoughts, emotions, decisions — all emerge from and reshape your local ripple configuration. Consciousness is not in the brain. It is ripple awareness interacting with structured vacuum.

Science is not wrong. It's misaligned.

Mainstream models are not failures — they are symptoms of deeper misinterpretation. We've been describing ripples as particles, forces, and flows. Now we describe them as what they are: tensions seeking permission.

UED is not just a theory. It is a lens.

It changes how we see machines, nature, life, society, and mind. It does not ask you to believe. It invites you to observe — differently, honestly, deeply.

Because when you stop looking for energy, and start sensing alignment — the universe finally speaks back.

Sandeep Chavan,

Independent Researcher, India

Don't miss out!

Visit the website below and you can sign up to receive emails whenever Sandeep Chavan publishes a new book. There's no charge and no obligation.

https://books2read.com/r/B-A-EPGPC-OIKUG

BOOKS 2 READ

Connecting independent readers to independent writers.

Did you love *Rethinking Energy: Ripple Tension, Not Substance*?
Then you should read *Our Story*[1] by Sandeep Chavan!

[2]

What if the universe was never separate from you—but trying to resolve itself through you?

Our Story is not a typical science book, nor a metaphysical manual, nor a philosophical treatise. It is a mirror—held up to the human experience. Drawing from the revolutionary framework of **Universal Energy Dynamics (UED)**, this book invites you to rethink everything: energy, truth, misalignment, consciousness, death, and even meaning itself.

In a world obsessed with particles, forces, and formulas, *Our Story* proposes a radical but intuitive idea: the universe doesn't operate on objects, but on **ripples**—tensions and alignments within

---

1. https://books2read.com/u/mdMZAO

2. https://books2read.com/u/mdMZAO

a structured field. And we, as conscious observers, are not merely in that field. We *are* that field remembering, resolving, evolving.

Across twelve deep yet beautifully accessible chapters, the book explores:

Why misalignment is not a flaw—but the birthplace of awareness.How consciousness is not given—but emerges as the field tries to feel itself.Why your thoughts, emotions, and even failures are ripple events—not personal faults.What happens when we stop chasing "truth" and start sensing alignment.How we can live *like the field*—soft, aware, aligned when needed, surrendered when not.

This is a story of us—not as static beings, but as dynamic, trembling possibilities.

Written by **Sandeep J. Chavan**, a bold thinker and independent researcher, *Our Story* brings the core principles of UED into everyday life—without equations, jargon, or dogma. It's for scientists who feel stuck, philosophers seeking simplicity, and anyone who's ever wondered, "Why do I feel so out of place in this universe?"

You'll leave this book not with more information, but with clearer perception—and perhaps, a gentler relationship with your own mind.

This is not the end of the story.It's just the permission to see yours differently.

Read more at https://www.sandeepjchavan.com/.

Also by Sandeep Chavan

A Monk on World Tour
The Making of the Monk

Chavan's UED research
CMGED Research Report
Chavan's Cosmic Continuum (C^3)
Frequency
Ripples of Reality
Reality Realigned
The Great Realignment
I – The Cosmic Leverage
Chavan's Cosmic Theory of Relativity (CTR)
Chavan's Theory of Light
Feeling Like a Physicist
TIME – The New Definition
$E \neq mc^2$: The Collapse of Classical Assumptions
The Chavan Codex
Chavan's Atomic Model
Chavan's Theory of Everything

Echoes of War: The Russia-Ukraine Conflict and Its Global Ripples

Shattered Lives
The War Economy
Borders Redrawn
Arms and Allies
Collateral Damage
Justice in the Ashes

Somewhere in the Darkness of the Cosmos

The Blueprint of the Universe
The Cosmic Breath
From the Edge of the Universe

The Eternal Sage Chronicles

The Sage and the Sovereign
The Family Pillar
The People's Flame
The Warrior's Soul
The Bridging Mentor
The Guiding Light

The Exodus

The Last Cycle: 2075

The Fundamental Theory of Life & Existence

Energy in Action: The Soul Demystified

The Active Mind

Energy in Connection

Death: The Disconnection

Subconsciousness: The Chosen Disconnection

Desires: The Engine of Life

The Restoration of Human Knowledge

Restoring Bhagavad Gita

The War of Thoughts

Cracks in the Known

The Human Game

The Divine Dilemma

The Original Intent

The Answers

Ultra Modern Life Philosophy

The Duality of Laughter and Tears

Take It Easy

Compare to Accept

Standalone

It's Not AI, It's AHI - Amplified Human Intelligence
The Decision Paradox
The Four Sapiens
Malicious Script of Indian Polity
The IIT Legacy & Global Impact
The IIT Legacy & Global Impact
You are just a Version of Your Original
Win the Game You Didn't Choose
Always Keep Your Bags Packed
Reveal to Shield - Learn the Game of Facades
Engage Beyond Elections
Culture, Identity & Change: The Evolution of Indian Society
The Triangular Dynamics
Never Trust Without Doubt
If You Can't Change the Country, Leave the Country
Live Fearless with Fear
India: The Investment Magnet
Realignment
Powering a Billion Dreams
Universal Skill
Extension: From Brush to Algorithm
Human Side of Einstein
The Future Knows No Lines
Let's Bring the Generations to the Table
Unified Energy Dynamics
A Beat That Never Dies
Strategic Horizons
Right From Scratch
The Forgotten Innovators
Physics Fixing
The Savior Illusion
The Selfish Side
The Age of the Weakest Sapiens

Neural Wars
The Absorption: There Is No Death
Strange Therapist
Bubbles: Become a Drop
Beyond the Blue
Half-Scholar
The Law of Truth
Failure
Philosophy 2.0: A Structural Manifesto
Chaos: A Gentle Guide to the Unresolvable
Self-Help: New Syllabus
Rippling
Game: How to Play the Share Market
Out of Syllabus
Saffronized: The Code Behind the Crown
The Anger
Moksha: The Liberation
Chavan vs Einstein: The Universe Was Never Curved
Rocket Science: The Ripple Gambling
The Cosmos: The Still Universe
The Taste of Silence
Chavan's Post-Materialist Human Model
10-Second Rule
Chavan vs Hawking
It was Never a Wave Signal
Quantum Courtroom: The Trial of a Century
Chavan–Heisenberg Ripple Uncertainty Principle
The Great Disruption
Dark ≠ Hidden
Breath
Calculus of Life
The Earth Doesn't Move
Black Hole Don't Exist

Chavan vs Kant

Naked Truth: Everything and Else

The Moon: Co-Author of Life's Script

World War-3

Aham Brahmasmi

Chavan's Declaration

Light: Neither a Wave, Nor a Photon

Chavan vs. Plato

Good Morning

Myth Wrapped in Math

Our Story

Rethinking Energy: Ripple Tension, Not Substance

Watch for more at https://www.sandeepjchavan.com/.

www.ingramcontent.com/pod-product-compliance
Lightning Source LLC
LaVergne TN
LVHW090308210926
848480LV00001B/169
9798231100590